HOW TO BUY
SOLAR HEATING
...without getting burnt!

HOW TO BUY
SOLAR HEATING
... without getting burnt!

Malcolm Wells/Irwin Spetgang

Illustrations by **Malcolm Wells**

 Rodale Press Emmaus, PA

Printed in the United States of America on recycled paper

Library of Congress Cataloging in Publication Data
Wells, Malcolm.
   How to buy solar heating without getting burnt!

   Bibliography: p.
   Includes index.
   1. Solar heating.   I. Spetgang, Irwin, joint author.   II. Title.
TH7413.W44   697'.78   77-27962
ISBN 0-87857-199-X

2 4 6 8 10 9 7 5 3

Contents

Chapter 1/ Welcome to the Exasperating and
Exciting World of Solar Heating 1

Chapter 2/ Basic Training 5

Chapter 3/ Your Best-Spent Dollars—Insulation 33

Chapter 4/ Is Your Home Suited? 79

Chapter 5/ Is Yours a Solar Family? 111

Chapter 6/ Our Poll—Solar Homeowners 141

Chapter 7/ Contracts and Contractors (and Architects) 159

Chapter 8/ Sunlight and the Law 189

Chapter 9/ The Sun Isn't Free 215

General Appendix 235
    Readers Checklist 235
    Manufacturers of Solar Collectors 238
    Solar Kit Manufacturers 246
    Solar Directories 247
    Federal Publications on Solar Energy 249

Index 257

Welcome to the Exasperating and Exciting World of Solar Heating

We asked them, "Does your solar system provide enough heat for your comfort?"

We asked them, "If a contractor did your solar installation, what was your greatest frustration in dealing with him?"

We asked them, "What would you do differently, if you had another opportunity?"

We didn't ask them: "Why did you go solar?" because each reply would have filled a book, rather than the questionnaire* we sent to most of America's solar homeowners. Answers to the "Why solar?" question

*A copy of our questionnaire is found at the end of chapter 6.

Homeowner's dream.

would have ranged from the homeowner's environment to his or her value judgments, dreams, and hopes.

Deciding on solar heating is one thing. Putting it into action is quite another. The industry is so new, almost everyone in it is still learning. And, as in every industry, there are a few less-than-reputable companies. This is not, however, to say there aren't perfectly acceptable and successful solar heating systems in use. There are. But getting there can be fraught with problems. And the best way to avoid those problems is to be tuned in, yourself.

You should know a great deal about the potential problems of a solar installation before you turn the job over to a contractor. That way, you won't be paying for *his* learning time. You should know enough to recognize the pitfalls in such areas as heat exchangers, thermal storage, heat-loss resistance ("R" values), sun rights, active and passive systems, etc. Sounds like a whole new language, doesn't it? Well, don't be discouraged. We've written this book to take you step-by-step through the whole process. In *How to Buy* . . . we offer our own solar experiences* and those of many solar home dwellers, salted with research, tips, and common sense. By the time you're through, you'll feel quite comfortable about how to buy solar heating without getting burnt.

A whole new language?

The answers on the questionnaires returned to us by America's solar homeowners revealed much about this burgeoning industry†

---

*We are the officers of a solar consulting firm dealing in feasibility studies, active (liquid and air) and passive systems, and educational publications and exhibits.

†Chapter 6, our solar poll, provides a digest of the survey results.

Their solar experience ranged from a few months in some cases, to ten, twelve, and eighteen years in others. Most of those who have had their systems for a long time installed them themselves, and many even designed them as well. Such people were the pioneers of the solar game. But, pioneers or newcomers, their efforts are still being closely watched.

One irate respondent writes, "Because of improper installation and design, I have filed suit against the contractor for fraud and misrepresentation."

Another says, "Our system employs manually operated dampers. If we were to do it over again we would incorporate dampers which would automatically close when the fan turns off."

"I show our home in conjunction with the Colorado State Solar School," says one solar homeowner, "to hundreds of people, since it is unique in construction, application, design, and function (100% solar heating)."

One from New Mexico relates, "We received a great many visitors during construction of the house, even visitors from foreign countries. These visitors lost us valuable time from construction, and for a while we seriously contemplated putting in a donation bucket."

And there was some humor. "The things I chuckle about are mainly the questions I get: 'You mean it works even in cloudy Oregon?' (Yes!) 'How much electricity does it make?' (It's a heating system, not an electrical generator.)"

And more: "I always got a kick out of comments of passersby while the house was under construction. People assumed I was only a laborer (I guess I looked too young to own a house) and were uninhibited in their comments. Most were positive, but I much preferred the comments about how ugly the house was. . . ."

So you see, it's not as simple as installing a conventional heating system. Also, the subject now seems to fascinate everyone, so be prepared to be the object of newspaper articles, visits from strangers, and curious stares. It will bring a whole new dimension into your life.

A young couple in New Jersey says, "Living in a solar home certainly makes one a sun worshipper. Man's relationship with the sun is more important than most people realize. In general, people take the sun for granted, but it is a fantastic star! Our dependence on it for heat has given us a new outlook on winter; most people in 'unsolar' homes look upon the beginning of winter, December 21, as a dead, 'nothing' period. Actually, from that day forward, we are collecting more solar heat each hour, and the days are getting longer as well. It's great. We love it!"

One solar homeowner wrote across the top of his questionnaire, "I hope you're serious about writing the book, because it's really needed."

We're serious.

2
Basic Training

A recent book for homeowners on solar heating launches the subject by offering this illustration. Its title: "A Simplified Diagram of Almost Any Solar Heating System." Not bad, eh? Just what you always wanted to know about solar heating . . . just what every homeowner needs in dealing with the energy crisis—another incomprehensible diagram.

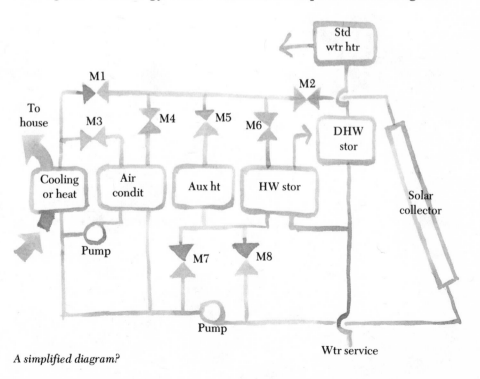

A simplified diagram?

Whew! Isn't that an awful-looking thing? Would you want one of them in *your* house? Do you see now why we wanted to write a simple, understandable book for people just getting into the solar world? That diagram is enough to make a person reject all thoughts of solar heating before he even starts. It's gobbledegook. It's an attempt, by some dumb technocrats, to illustrate a phenomenon as simple as a sunbath.

Look at those symbols! What do you suppose the letter *M* next to a bow tie (⋈) represents? Motor? Meter? Manual?

Would you believe "valve"?

That's right; each of those little bow ties represents a cutoff point in the system. But even *that* knowledge does you no good, because there's little indication on the diagram as to which way most of the water (or is it air?) is flowing, or of what is supposed to be happening there.

And what do you suppose the letters *DHW* stand for? I have no idea. D-something hot water, I suppose. *Direct* hot water? Aha! (💡) *Domestic* hot water! Of course. All it took was a little concentration. But why didn't they just *say* "domestic hot water"? Why do they make us go all through that guessing-game business when it's so easy to illustrate a typical solar system? Look:

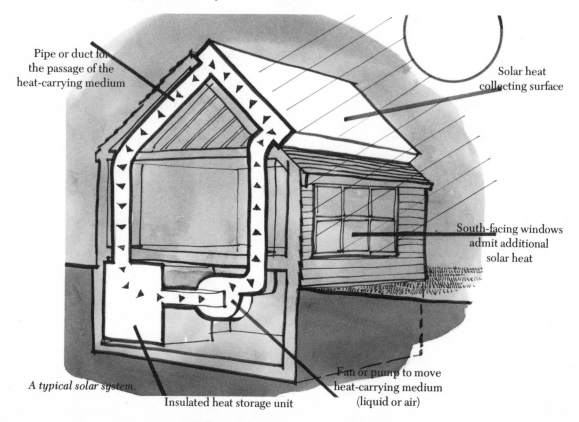

Pipe or duct for the passage of the heat-carrying medium

Solar heat collecting surface

South-facing windows admit additional solar heat

A typical solar system.

Insulated heat storage unit

Fan or pump to move heat-carrying medium (liquid or air)

Solar pioneer Harry Thomason,* whose patented "Solaris" system is still one of the simplest, least expensive, and most efficient on the market, likes to tell about the way he discovered the sun's great ability to heat water. During a brief summer rainstorm he took cover beneath a rusty old corrugated iron roof, and when he stepped out into the sunshine a few minutes later, he felt *hot water* dripping from the eaves.

That's what solar heating is all about.

After his discovery, all Dr. Thomason had to do was make sure the same phenomenon could be made to occur in wintertime, and he was on his way to the patent office. The simplicity of discoveries and inventions of this type is a good excuse for optimism about the future. But we'll all have to be watchful if we want to keep solar heating simple enough to stay under individual control. Sunlight is universal, dependable, and free, but there are people who would like nothing better than to make solar heating seem so arcane and complicated that we amateurs would be allowed to use it only under the control of a utility company.

Solar scientists and engineers routinely use words like *delta-T, U-factor, K-value, langley, micron,* and *selective surface,* the last of which sounds vaguely like the name of your local draft board; all of which are enough to scare you away from the whole business. But there's good news: you don't need to know any of those words.

Few of us understand what really goes on under the hoods of cars, yet each of us drives thousands of miles a year without having breakdowns on the road. As drivers, we learn to recognize the basic warning signals, and in most cases can arrange for necessary repairs before highway emergencies occur. (Most accidents, remember, are caused by driver error and not by mechanical failure.)

Solar heating, many times simpler and infinitely safer than the operation of a motor vehicle, need not intimidate us at all; many of the warning signals we must recognize in order to handle it successfully can be found right here in this book. Solar heating is basically so simple, in fact, that one of this book's authors—the architect—has designed several successfully solar-heated buildings without ever having grasped the meaning of *delta-T, U-factor,* or any of those other words. We mention this not because such words are unimportant—they are important; after all, someone has to understand them—but rather to underscore the fact that the average homeowner need not know them in order to enter

*Dr. Thomason's address and those of most individuals who are key to the solar industry can be found in Dr. Shurcliff's *Solar-Heated Buildings–A Brief Survey,* at the end of chapter 6.

negotiations with solar heating contractors and suppliers. We'll cover that subject—negotiations and contracts—more fully further on in this book. Right now it's time to take a look at the world energy picture and see where we're going . . . see how solar energy is involved in our future.

The three big energy users today are—

    industry-agriculture,
    transportation, and
    residential-commercial,

each taking roughly a third of all the energy produced by America's big energy companies. Notice that whenever you see such figures as these, the huge impact of *sunlight* in raising food, growing cotton, producing wood, drying clothes, and *heating our buildings for most of the year* is never shown. The power companies don't know how to deal with an energy source that's free and available to everyone, so they try to pretend it doesn't exist.

The world's supplies of oil, gas, and coal do have foreseeable limits. We know that nuclear power is expensive and awesomely dangerous. We hear that another and presumably safer kind of nuclear energy called *fusion* is at least a generation away. And we're learning that natural forces like waterfalls, tides, and earth-heat are, while promising, all too often inconveniently located and unable to meet today's and tomorrow's huge energy demands. Other energy sources, such as the combustion of wood, of sewage, and of garbage, and the managed decomposition (bioconversion) of wastes are also promising, but they're scaled to a level of energy-use we seem at least temporarily to have abandoned. We can escape to Maine or Idaho and pretend that our low-energy world is the answer to everything, but the fact of Gary and Detroit, Phoenix and Pittsburgh, Birmingham and Newark, remains. If we tried to run our wasteful American industrial society on the burning of wood we'd make a treeless waste of this continent in no time.

What it all means is that if oil, coal, and gas are needed to run our industries for another generation or so while they convert to sun power, we'd better start using that sunlight for home heating just as soon as possible in order to save enough fuel for the transition period.

There's no question that we'll have to return to a less energy-intensive way of life. This drunkenness has got to end. The warning signs are everywhere. And sunlight is everywhere, too: dependable, worldwide, and pollution-free. Even on the cloudiest day the daylight you see is sunlight. It never fails.

Your house got most of its heat from the sun last year. That's an energy fact—and an economic fact—too often overlooked. We live in a solar energy society right now. This world we love would be an icy desert without the life-giving warmth of the friendly star. The heat we get from oil, coal, and gas is only a fringe benefit, a supplement, to the massive amount of radiation that even in winter keeps us safe from the unimaginable cold of a sunless planet. Both August's heat and February's cold lie within that narrow temperature band needed for life here to exist. Furnaces and fireplaces are recent, temporary, and relatively minor refinements to the solar system that's been heating the world for five to ten billion years.

Physiologically, we were designed to function without any heat other than that provided by the sun, both in winter and in summer. Food, clothing, and shelter are all we need to live through any weather. We evolved in ice caves and deserts. We are tougher than we think. But for the last few years—how many? one hundred and fifty?—we've stepped out of our earth-role and burned up most of the fuels. Now we must step back into that role and live on sunlight again.

The nice part is that no suffering, or even discomfort, is going to be involved. We've learned enough, during these wasteful years, to take the sting out of our return to a solar life. It's no longer going to be a matter of huddling together inside caves* all winter. Instead, the prospect is one of comfortable temperatures, clean skies, and much lower fuel bills. The reason: insulation—it makes all the difference.

We'll get into the subject of insulation in chapter 3. Right now, it's time to look at some of the solar systems that are in use today. There are literally hundreds of them, all kinds of systems at all levels of complexity, but the most popular and the most widely used are of just three basic types: air, liquid, and passive.

In *every solar heating system* the principle is the same:

1. sunlight heats a collecting surface;
2. that surface-heat is stored somewhere; and
3. the stored heat is used when it's needed.

In the illustrations that follow, you'll see this three-step principle as

---

*Although we'll do well to remember that earth-covered, solar-heated homes often need no supplemental heat at all.

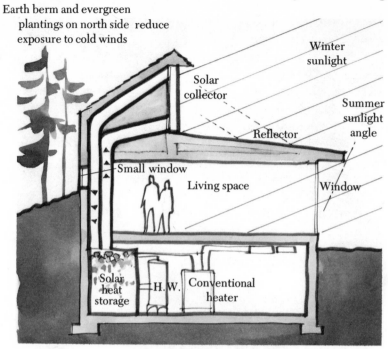

Earth berm and evergreen
   plantings on north side reduce
   exposure to cold winds

Winter
sunlight

Solar
collector

Summer
sunlight
angle

Reflector

Small window   Living space

Window

Solar
heat
storage   H.W.   Conventional
                      heater

An air system mounted on the roof (new construction).

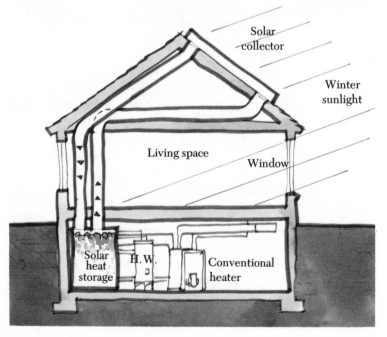

Solar
collector

Winter
sunlight

Living space

Window

Solar
heat
storage   H.W.   Conventional
                      heater

An air system mounted on the roof (new or existing construction).

it is expressed in the most common types of solar installations. Each type is shown in both new and retrofitted construction.

AN AIR SYSTEM MOUNTED ON THE ROOF

Air systems have the advantage of being simpler and somewhat more trouble-free than liquid systems today. With no water to freeze, or leak, and with no antifreeze to pose a contamination threat to drinking water supplies, air systems are the choice of many people, in spite of the fact that the rock bin needed for the storage of airborne heat must have three times the volume of a water-heat storage tank. But that figure is somewhat misleading. It isn't until you remember that a 3-foot cube has more than *three* times the volume of a 2-foot cube that you realize that three times the volume does not mean three times the length or width . . . or height.

The principal complaint voiced by the owners of air systems is the leakage of air from ducts and storage areas, allowing hard-won solar heat to escape, often back up into the collector, where it is lost to the wintry sky. Air systems have other problems, too, of course: problems such as their failure to provide high-enough domestic water temperatures; but these seem to be more than offset by the fact that, for space heating, air can be effective at much lower temperatures than can water.

Now, before we look at other systems, let's stop for a minute to see what percentage of your total heating job solar heat is likely to provide. That percentage depends upon a lot of things we'll discuss in chapter 4, but assuming that your site and your building are favorably oriented and exposed, it then comes down to a matter of where in the world you happen to live. If you can find your home region on the map on the next page you can see what the solar prospects are for that area. Remember that you can do a lot worse than the predicted percentages if you're not careful, and that you can sometimes do much, much better by using well-thought-out designs and good insulation.

CLIMATIC ZONES OF THE UNITED STATES

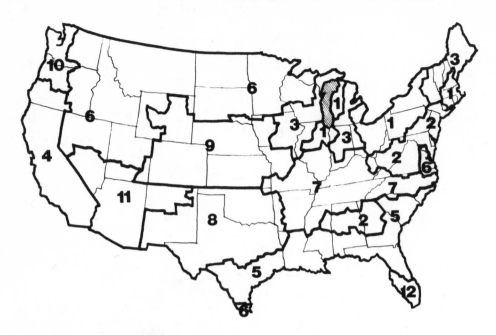

Breakdown of zones by state. Numbers in parentheses following each state
indicate zones included in that state.

Alabama (2, 5, 7)	Maine (3)	Ohio (1, 2, 3, 7)
Arizona (11)	Maryland (2)	Oklahoma (7, 8)
Arkansas (7, 8)	Massachusetts (3)	Oregon (4, 6, 10)
California (4)	Michigan (1, 3)	Pennsylvania (1, 2)
Colorado (9)	Minnesota (3, 6)	Rhode Island (1)
Connecticut (1, 2)	Mississippi (2, 5, 7)	South Carolina (5, 7)
Delaware (2)	Missouri (3, 7, 9)	South Dakota (6, 9)
Florida (5, 12)	Montana (6)	Tennessee (7)
Georgia (2, 5, 7)	Nebraska (9)	Texas (5, 8)
Idaho (6)	Nevada (6, 11)	Utah (6, 9, 11)
Illinois (1, 3, 7)	New Hampshire (1, 3)	Vermont (1, 3)
Indiana (1, 3, 7)	New Jersey (2)	Virginia (2, 5, 7)
Iowa (3, 6, 9)	New Mexico (8, 11)	Washington (6, 10)
Kansas (7, 9)	New York (1, 2, 3)	Washington, D.C. (2)
Kentucky (2, 7)	North Carolina (2, 5, 7)	West Virginia (1, 2, 7)
Louisiana (5, 8)	North Dakota (6)	Wisconsin (3, 6)
		Wyoming (6, 9)

Percent of Energy Supplied by Solar Heating System

Climatic Zone	Percent
1	71
2	72
3	66
4	73
5	75
6	70
7	70
8	71
9	72
10	58
11	85
12*	85

*Includes only hot water needs.

Now we'll go back to our review of the basic system-types, the next of which is:

AN AIR SYSTEM MOUNTED ON THE WALL

If you have the wall space to spare, and it's facing in the right direction (say, within 10° or 15° of due south), this design offers you a simple way to install a collector. Sheltered by a roof overhang in many cases, the collector on a wall is protected from rain leaks and from much

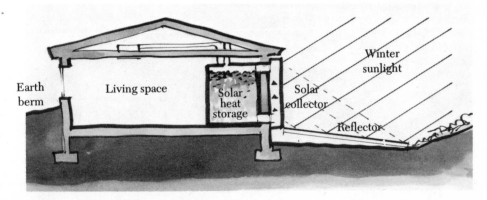

An air system mounted on the wall (new construction).

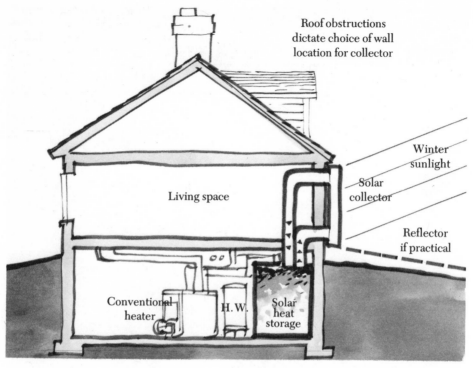

Roof obstructions
dictate choice of wall
location for collector

Winter
sunlight

Solar
collector

Reflector
if practical

Living space

Conventional
heater

H. W.

Solar
heat
storage

An air system mounted on the wall (new or existing construction).

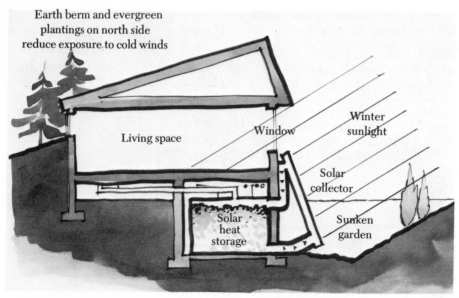

Earth berm and evergreen
plantings on north side
reduce exposure to cold winds

Winter
sunlight

Living space

Window

Solar
collector

Solar
heat
storage

Sunken
garden

An air system mounted on the wall (existing construction).

of the ice and snow damage possible with rooftop installations. Further, the roof itself is free of the potentially troublesome penetrations needed for the air ducts . . . and the collector can be shaded from the searing, unwanted heat of the high-angled summer sun.

Remember that we're showing, in each example, only a few of the hundreds of designs possible. You might think, at first, that the top and bottom air ducts to an air collector would limit its design to a single configuration, but just look at some of the ways in which the basic panel can be rearranged to offer duct-location flexibility!

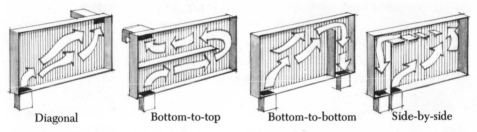

Diagonal   Bottom-to-top   Bottom-to-bottom   Side-by-side

Duct location flexibility.

The value of reflective surfaces used in conjunction with any solar system cannot be overstressed. If you can cope with the kid, or dog, or vandal threats to a reflector, you can slash the size requirements of your collector for the price of some shiny aluminum. Look:

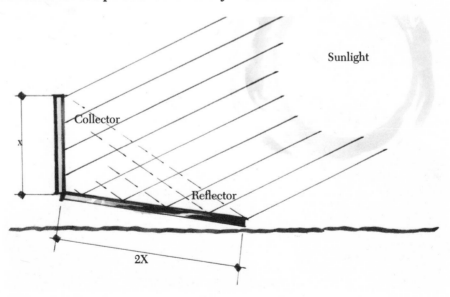

Reflector size rule of thumb.

A reflector-size rule of thumb for many temperate latitudes is two times as wide as the collector height. You get almost twice the energy for little more than one times the collector cost! The reflectors, if properly designed, can even be picked up and stored conveniently during the summer when they're not needed.

A WATER SYSTEM MOUNTED ON THE ROOF

You can see at a glance that this type of system is exactly the same, in principle, as the air systems already illustrated. The main physical difference, obviously, is that, because of water's three-times-better heat storage capability, many of the components have become smaller. As we mentioned earlier, the big problem here seems to be leakage and the possible contamination of drinking water by antifreeze or other collector-related nasties. The causes of such problems can range from poor design to poor materials and workmanship. Be sure to review this subject thoroughly before you sign that piece of paper the contractor gives you. A review of both our checklist and our summary of the national solar homeowner's poll will be a good preparation for such

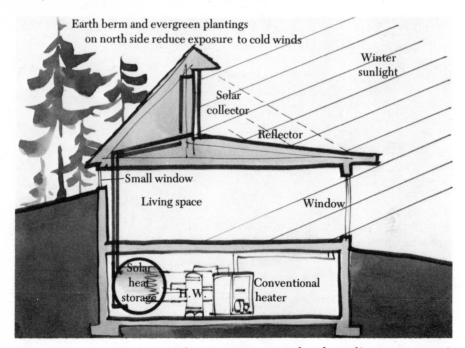

A water system mounted on the roof (new construction).

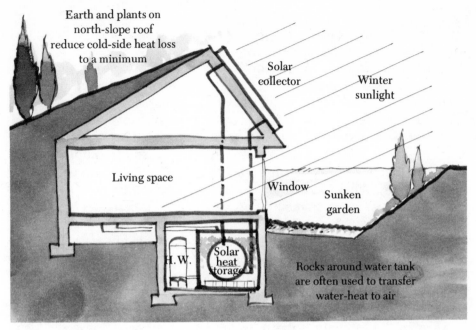

Earth and plants on
north-slope roof
reduce cold-side heat loss
to a minimum

Solar
collector

Winter
sunlight

Living space

Window

Sunken
garden

H.W.

Solar
heat
storage

Rocks around water tank
are often used to transfer
water-heat to air

A water system mounted on the roof (new construction).

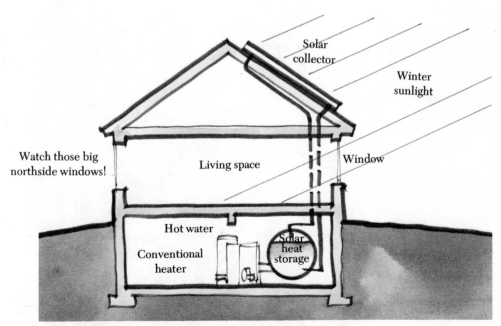

Solar
collector

Winter
sunlight

Watch those big
northside windows!

Living space

Window

Hot water

Conventional
heater

Solar
heat
storage

A water system mounted on the roof (new or existing construction).

discussions. Maybe as a result of all this you'll find a way to manufacture giant thermos bottles for storing solar-heating water. Who knows where this technology will lead us?

Many *existing* homes may not prove to be particularly suitable for solar space heating, though. A review of chapters 4 and 5 will help you quickly establish whether or not your present home is a good match for solar. Fortunately, most homes that are not suitable for solar *space heating*, do provide a fine starting point for solar *domestic water heating*.

The reason is that the size or capacity of solar components for domestic water heating is only about 20% of that required for solar space heating. Where you might need room to mount 500 square feet of collectors to heat your home (chapter 4 will discuss size of components vs. area to be heated in greater detail), you would require less than 100 square feet of collectors to provide for your domestic hot water needs. In a similar manner, other components—such as the solar-heated water storage tank—are also much smaller, making the installation in many homes that much easier.

The simplicity of a solar heating system for domestic hot water can be seen in the following photo. It shows a demonstration solar water

The solar exhibit designed by the authors at Philadelphia's Franklin Institute Museum. This is the first functioning solar energy heating system to be installed in a museum of national prominence, permitting the museum visitor to experience the sun's warmth as he pushes a button, operating the system.

heating system in the Futures Exhibit at the Franklin Institute, Philadelphia, that we designed and installed for them early in 1976. To our knowledge, this was the first functioning solar water heating exhibit to be shown by an internationally prominent scientific institution. The plumbing and interconnections are deliberately left uncovered so that the museum visitors can see the workings of the system as they press a button, activating the fan behind the heat exchanger.

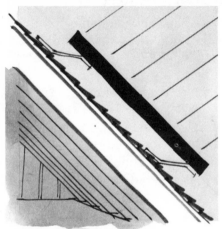

Collector supported by galvanized steel plates bent to proper shape and carefully flashed and sealed at roof.

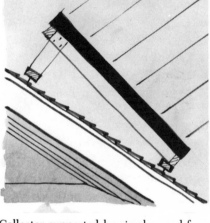

Collector supported by simple wood framing . . . NOTE roof mounts raised on "pins" to avoid trapping water and ice on roof.

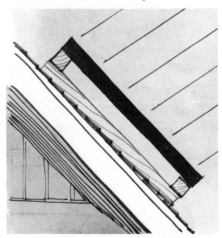

Collector supported by heavy timbers arranged to allow free drainage of water from roof.

Rooftop mountings.

Collector becomes a part of the roof. NOTE head and foot flashing, and remember to allow for expansion and contraction of unit in all flashing around and between collector units.

For homes with only limited suitability for solar space heating, a simple (and inexpensive) domestic water heating system of this type could be the best solution. Now, back to water systems mounted on the roof.

Some rooftop collectors are mounted directly on roofs, becoming, in effect, the roofs themselves. Other collectors are mounted on brackets or frames, up off the roof surface as shown here in order to interfere as little as possible with the roof's main job. This is another detail you'll want to look into carefully before you take your big solar step. Involved here are such considerations as the insulation of the back of the collector (tight-on-the-roof collectors lose their back-heat into the building that needs it), and the sealing and flashing of the brackets, pipes, ducts, and other features which must mount on, or penetrate, the roof. Great care is the key to success.

Here's a table showing the approximate collector and storage tank sizes you'll need for solar-heating a house of about 1,500 square feet. First you must refer to the regional key map here. Then you can find the figures for your region, and adjust them as needed if the heated floor area of your house is something other than 1,500 square feet. Remember that it's not just the size of your house, or its insulation quality, or even

Collector and Storage Tank Sizes Required for a 1,500-Square-Foot Home

Climatic Zone	Collector Area (sq. ft.)	Storage Tank (gal.)
1	800	1,500
2	500	750
3	800	1,500
4	300	500
5	200	280
6	750	1,500
7	500	750
8	200	280
9	600	1,000
10	500	750
11	200	280
12*	45	80

*Includes only hot water needs.

its amount of sun exposure so much as the way you use your house. Tests have shown that identical side-by-side buildings can have heating bills that vary by as much as 500%. You'll have to decide at some point whether to rate yourself as an energy miser, an energy waster, or someone in between.

A WATER SYSTEM MOUNTED ON THE WALL

There's little else that need be said about this one, since it combines all the applicable features of systems already described. It's nice if you can do it, particularly if you can arrange a reflector successfully, but very few houses have the unbroken south-facing walls needed to mount these panels.

Most solar designers use as a rule of thumb, for determining the best winter-tilt angle for collectors, 10° or 15° added to the local latitude as the angle (above horizontal) for the collectors. At the Philadelphia/Indianapolis/Denver/Salt Lake latitude the collectors should slope a little more than 50° above horizontal. So where does that leave these vertical, wall-mounted collectors? Thirty-five degrees off base, that's where. A reflector helps a lot, but you must remember that a 35° turn away from the ideal means an 18% loss of energy. Still, that's a lot better than

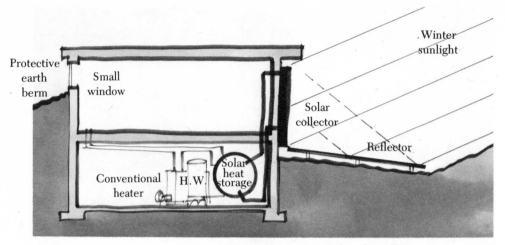

A water system mounted on the wall (new construction).

mounting the collectors flat on a flat roof. Then, in the latitude under discussion, they'd be a full 55° away from the optimum angle and would then lose 57% of the available energy. Like everything else, solar heating design comes down to a matter of compromise; you give up this in order to get that.

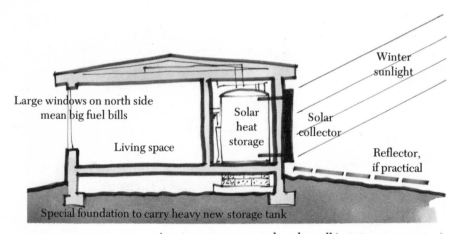

A water system mounted on the wall (existing construction).

A SOLAR SYSTEM NEAR A HOUSE

If there is just no way you can get a collector onto your roof or wall, or if your house doesn't face the right direction, you can perhaps

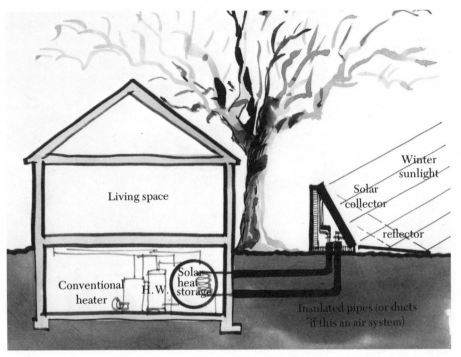

A solar system near a house.

still heat it successfully by installing an outdoor solar unit a short distance away. Such collectors are already on the market and are basically sound in concept, although we have heard from the owners of such systems that they were undersized and overpriced. Competition, however, is already taking care of that situation. Knowing what you already know about the probable collector size needed in your part of the country, you can be confident of your ability to assess the systems that will be proposed to you. By the time you finish this book you'll be able to spot the gyps almost on sight. When considering a remote system be sure to keep in mind all the additional things that can happen to a big mechanical thing sitting on icy soil in the middle of a February night.

PASSIVE SYSTEMS

A friend of ours has a black-painted wood door on the south side of his house. Outside the door, there is, as there should be, a tight-fitting glass storm door. When he opens the black door from the inside, on sunny winter days, a rush of hot air spills into the room. He has a passive solar collector. Perhaps the simplest passive collector of all is a south-facing window. The phenomenon of sun heat on winter days is

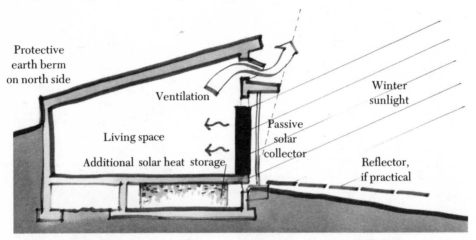

A passive system on a wall (new construction).

familiar to us all: pleasant under most conditions, unpleasant when overheating occurs. Most passive solar collectors rely on their great mass to absorb the excess daytime sunlight which, coming through an ordinary window, could cause such overheating. These collectors have many variations and refinements. Some are simply great slabs of concrete painted black, behind glass, slabs which react so slowly they don't start reradiating their solar heat to the rooms behind them until after the sun has set. Others have top and bottom air vents to help speed the heat to the rooms. Still others have insulating devices that prevent the slabs from losing too much heat back out through the glass at night.

"Passive" is a hot word in solar circles these days, and of course the concept *is* appealing—imagine: no motors, no pumps, no parts to wear out!—but in our interviews with passive system owners, we've found erratic behavior to be the most common complaint. We believe that perhaps some combination of passive and active (mechanical) solar systems may be ideal; using as little motor-energy as possible but having some means to control the temperatures before they get too far out of hand. It's no fun to sit up all night with an overheated slab . . . or a frigid one.

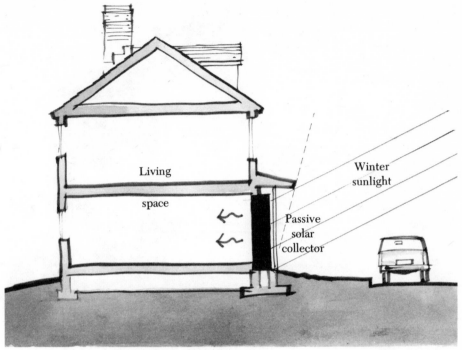

A passive system on a wall (existing construction).

Here again, it's good to remember how much more efficient water is as a heat-storage medium than are rocks or concrete. Some passive systems make use of hundreds of freeze-proof (plastic) 1-gallon jugs* full of water, instead of a concrete wall. Others use the two in combination. And there is on the horizon a new automatic passive water system that uses temperature-activated oil "valves" to move the sunny water to the inside of the wall for the night.

You can increase the passive solar effect at your house simply by adding more south-facing glass . . . skylights, windows, greenhouses . . . those kinds of things. Not only you but your houseplants, too, will be delighted by all the added brightness and warmth. As long as you have a way to store all the incoming heat, the more the better, at least in the cooler parts of the country. There's probably a limit to the number of plastic water jugs and 55-gallon drums you'll want to have standing

*See our poll results in chapter 6 for the hazards of plastic jugs.

Interior finish and exterior siding removed from south wall studs, then two layers of fiberglass sheeting are installed and carefully sealed at all edges. Result: a very low-cost passive solar collector which, if insulated at night, can contribute a considerable part of the house's wintertime heat. Note use of exterior reflector for added energy gain.

Increasing the passive solar effect (existing construction).

among your antique Chippendales, but that limit will vary in proportion to the amount of your fuel bills, so don't make your final decision in July.

The only other catch in the direct-sunlight idea, and we might as well repeat it right here, is that *unless you insulate the glass at night and during cold, cloudy weather, your house will lose all the heat the glass gained, and then some.* Another reason to provide such glass with insulating covers is that of summertime overheating. Even in the spring and fall, and occasionally even in winter when a freak 80°F. (27°C.) heat wave passes through, you'll want to be able to shut out the unwanted radiation from the sun. Solar radiation, of course, can be shut *out* by nothing more substantial than a flimsy sheet of mirrored Mylar, but mirrors won't keep heat *in*. For cold-weather insulation you must use a thick layer of insulation that fits tightly at the edges. If it fits tightly to the glass as well, so much the better, although in practice this is very

hard to arrange, and can even cause additional window-cleaning problems. The owners of an old house in Philadelphia simply stick bubbly,
clear plastic sheets—the kind used to protect things like cameras during
shipment—directly to the windowpanes each winter, and thereby
reduce drafts to a minimum. We use—and recommend—either folding
or sliding panels of insulation board surfaced with any material that
will make them durable, attractive, and fireproof . . . indoors, of
course; few of us are dedicated enough to close outdoor shutters every
time the sun disappears. The urethane-board panels of the shutters in a
New Jersey office building are surfaced with a thin coat of supertough,
fiberglass-shred-reinforced plaster. As little as $1/16$ inch can be troweled
onto any roughened board, successfully, by anyone. The material is
called Surewall by the J. R. Bonsall Corporation of Lilesville, NC 28091,
and Blocbond by Owens-Corning of Fiberglas Tower, Toledo OH 43659.

Zomeworks Corporation (Box 712, Albuquerque, NM 87103) sells
several types of manual, heat-activated, and electrically operated
devices for the insulation of glass openings in roofs and walls. An
avalanche of these and similar materials seems ready to burst upon the
nation, so be ready to take advantage of them. Take advantage, too, of
the good books (See Bibliography) now available on the subjects of solar
greenhouses, skylights, and window-mounted solar devices. Every
dollar you spend on such things is likely to save you another dollar—or
more—in fuel bills.

Notice that in each of the examples we've reviewed, exactly the
same things happened:

1. sunlight heated a collecting surface;
2. that surface-heat was stored somewhere; and
3. the stored heat was used when it was needed.

Solar heat, of course, can be used for room heating alone, for both room
heating and domestic water heating, or for domestic water heating
alone.

Since the subject of this book is solar *heat*, we're going to pass very
quickly over two closely related subjects, solar air conditioning and
solar electricity.

Solar air conditioning requires high water temperatures, and that
generally implies the use of focusing or sun-tracking (moving) collectors, of greater safety precautions, and of more sophisticated heat-
storage facilities, all of which mean money. Solar air conditioning for

homes, particularly for our poorly oriented, poorly summer-shaded existing homes, seems to be impractical in all but the hottest parts of the United States, so you would be well advised to seek competent solar engineering help before getting committed to the installation of any solar cooling project.

The prospects for solar air conditioning in the future, even for the more temperate zones, are bright. A lot of people are already at work on this problem, and there are already indications that solar air conditioning using collected temperatures as low as 160°F. (71°C.) will be available within a decade. This means that present-day collectors won't have to be replaced when the new cooling technology is available.

Solar electricity, generated directly by means of solar cells mounted on roofs and walls, is still many times too expensive to compete with utility-company-generated electricity, but it, too, offers great promise for the future. Imagine having a silent, nonpolluting electric car that got recharged each night by nuzzling the stored sun energy that fell on your roof that day! It's a likely prospect, and it relates closely to two other aspects of the solar-electric subject. One is that of wind-generated electrical power; the other is the storage of such energy.

It is rather astonishing, when you think about it, to realize that *every watt of electricity you use is being generated at the very instant you're using it.* None of it has been stored, at least not in the form of electricity. That's why utility companies talk so much about leveling out the demand. They'd need far less generating capacity if we could use, or store, at 3:00 A.M. and on April Sundays some of the electricity we use in such great quantities at peak hours during the air-conditioning or the electric-heating season. Energy storage is a big consideration in any system, and it seems most difficult to solve with respect to electrical energy. The stuff's so slippery and so fast it won't stay on the shelf at all. That's why anything we can do to use the off-peak excess can help the national energy picture. You might investigate the off-peak rates in your area and, if they're promising, think about heating and storing water (or the more sophisticated phase-change salts) for use when the rates are high.

At the moment, most home-generated electricity is produced by wind power, which is, of course, simply sun power one step removed. The sun makes the wind blow. And almost all wind-generated energy is stored in rows of ordinary car batteries, standing on shelves. They're fairly expensive and heavy, and they require a lot of care and ventilation. And they wear out. But promising developments in hydrogen technology point to a safe and trouble-free kind of storage that's compatible with home generating systems.

Will all these developments tend to make today's solar heating obsolete? It's a question we're asked all the time, and our answer, based on years of experience, is no. Home wind systems, batteries, hydrogen systems, and solar cells all tend to be equipment we'll *add to* our solar heating system, not things that will replace it. There is, after all, only so much energy in a certain size sunbeam. Future equipment may be somewhat more efficient in collecting and storing that energy, but the practical limits are already within sight. We may someday find it advantageous to replace our solar components with more efficient devices but the basic requirements of location and slope will never change.

If we come as close as possible to ideal sun angles and orientation the first time we install a solar device no future development can make them less ideal. The sun track across each local sky is fixed forever; no more and no less sunshine will ever reach that spot.

Well, that's not completely true. As we move away from the practice of burning fuel to get heat and power, the smoke that now obscures every horizon will slowly disappear, allowing more sunlight to penetrate the atmosphere, thereby increasing the efficiency of all solar systems, particularly those in the most heavily industrialized states.

What that change will do for the human spirit can hardly be imagined. The lift we get whenever sparkling clean arctic air arrives seems to forecast a future we thought existed only on calendars and postcards.

Now that you're in such an exalted mood, let us bring you down to earth. It would be unfair to end a chapter on solar basics without a word or two about how much your solar heating system is likely to cost. The costs are bound to sound high to you at first, and maybe they are, but remember that we'll have a lot of comforting things to say about life-cycle costs, about paybacks, and about energy financing in chapter 9.

Don't lose heart. You can do it if you care enough.

Range of Costs per Square Foot of Home Living Space

Climatic Zone	Cost Range (per sq. ft.)
1	$5.30–$16.00
2	$3.30–$10.00
3	$5.30–$16.00
4	$2.00–$ 6.00
5	$1.30–$ 4.00
6	$5.00–$15.00
7	$3.30–$10.00
8	$1.30–$ 4.00
9	$4.00–$12.00
10	$3.30–$10.00
11	$1.30–$ 4.00
12	$ .30–$ .90

APPENDIX / CHAPTER 2

BIBLIOGRAPHY

Buy Wise Guide to Solar Heat. Research and compiled by Floyd Hickock, Hour House, St. Petersburg, Fla.
A technology transfer report—assembling many inputs from various experts on the individual factors involved with buying solar.

Climate and Site: Influence on Passive Solar Building Design. Michael Holtz from AIA Research Corp., 1735 New York Ave. NW, Washington, DC 20006.

Design with Climate—Bioclimatic Approach to Architectural Regionalism. Victor Olgyay. Princeton: Princeton University Press, 1963.

Direct Use of the Sun's Energy. Farrington Daniels. New York: Ballantine Books, 1964, by arrangement with Yale University Press.

The Food and Heat-Producing Solar Greenhouse. Rick Fisher and Bill Yanda. John Muir, P.O. Box 613, Santa Fe, NM 87501. Distributed by Bookpeople, 2940 7th St., Berkeley, CA 94710. 1976.

In the Bank . . . or Up the Chimney? A Dollars and Cents Guide to Energy-Saving Home Improvements. 1975. Available from the Superintendent of Documents, U.S. Government Printing Office, Washington, DC 20402; order number 023-000-00297-3; $1.85*
A consumer's guide to energy conservation.

Informal Directory of the Organizations and People Involved in the Solar Heating of Buildings. Available from William A. Shurcliff, 19 Appleton St., Cambridge, MA 02138.

Proceedings of the Solar Energy-Food and Fuel Workshop. Edited by M. H. Jensen, Environmental Research Lab, University of Arizona, with ERDA and USDA, April 1976.

Producing Your Own Power. Edited by Carol H. Stoner. Emmaus, Pa.: Rodale Press, 1974.

Research Evaluation of a System of Natural Architecture. Ken Haggard, on Harold Hay's Atascadero, California, Skytherm House. Available from National Technical Information Service, Springfield, VA 22161; order number PB 243–498; $10.50.*

*Please contact source for latest price before ordering as prices subject to change.

Save Energy: Save Money. Published by the Institute on Energy Conservation and
the Poor, Office of Economic Opportunity, Washington, D.C. Single copies
available from the Community Services Administration, 1200 19th Street,
N.W., Washington, DC 20032.
Energy conservation and do-it-yourself solar heating equipment for low-income
people.

Simulation Analysis of Passive Solar-Heated Buildings. (LA-UR-76-89). Doug
Balcomb and Jim Hedstrom. Los Alamos Scientific Lab, Solar Energy Lab, Mail
Stop 571, Los Alamos, NM 87544.

Solar Energy. William W. Eaton for ERDA, Office of Public Affairs. Available from
Dept. of Energy, 20 Moss Ave., N.W., Washington, DC 20545.

Solar Energy for Space Heating and Hot Water. Division of Solar Energy, ERDA. SE
101. Available from the Superintendent of Documents, U.S. Government Printing Office, Washington, DC 20402; order number 060-000-00006-9; $.35.*

Solar Energy and Your Home. Jan. 1977, HUD-PDR-183 (3). Available from the
Solar Information Center, P.O. Box 1607, Rockville, MD 20850.

The Solar Greenhouse Book. Edited by James C. McCullagh. Emmaus, Pa.: Rodale
Press, 1978.

Solar Heating Papers. Available from the author: Norman Saunders, P.E., 15 Ellis
Rd., Weston, MA 02193; guide: $.50; basics: $4.00; supplement: $4.00; builder's
set: $4.00; index: $1.00.*
Emphasizes passive systems and energy conservation. Full set comprises a self-
teaching course from high school level through construction and engineering.

Sun/Earth. Solar Group/Architects, Crowther/Solar Group, 310 Steele St., Denver,
CO 80206. 1976.

*Please contact source for latest price before ordering as prices subject to change.

3

Your Best-Spent
Dollars—Insulation

Let's talk about insulation.

Do you think your house has enough of it? If it does, you are either a rare bird . . . or you're mistaken. Fewer than 2% of the private homes in America are well insulated. Some are what might be called adequately insulated, but many have almost no insulation at all.

YOUR FIRST LOOK; A COMMONSENSE APPROACH

Think of your home as a container with openings. The container is composed of walls, ceiling, and floor, while the openings are primarily windows and doors, cut into the walls of the container. The most obvious insulation needs of this container relate to air leaks. Although these leaks may not have the single greatest impact on your energy usage, they are usually the most obvious because they make some portions of your house uncomfortable (and in some cases unusable) in extreme weather conditions.

Start by thinking of your house as this container we are discussing. What rooms are most comfortable in extreme weather? What rooms least? Concentrate on the least comfortable ones. Are there any obvious air leaks? Check around the windows and doors. Is there a fireplace with an open flue, drawing your room's warmth to the outside? Do the windows fit tightly . . . or are there gaps around the framing, or between window-sash members? Is your window air conditioner well seated, or is it a source of cold drafts in the winter? If the room has a

Think of your home as a container with openings.

Which house has ceiling insulation?

door leading to the outside, is it snug or can you feel air leaks around its edges or framing?

How about your roof after a snowfall? If similar homes in your neighborhood accumulate snow on their roofs more quickly than yours, and if yours is the first to melt clean, it's a sure sign of trouble. It says your costly heating energy is escaping through your ceilings, and warming your roof, not the inside of your home.

We could go on and on with examples of obvious telltale signs of poor home insulation. The important thing, though, is that you understand the commonsense approach to defining your home's insulation needs. *Stand back, observe, and really think about what you see and feel.*

Part of the commonsense approach to minimizing energy-use relates to conservation, as well as insulation. Some of the easiest conservation actions you can take (at little or no cost) are:

- Lowering your thermostat setting in winter, and raising it in summer. The "comfort range" of 68° to 76°F. (20° to 24°C.) established years ago *is a myth.* Unfortunately, generations of Americans have been raised to believe without question that they're uncomfortable in winter temperatures below 68°F. (20°C.). After three years of gradually declining thermostat settings, many families find that a daytime setting of 64°F. (18°C.), and a nighttime setting of 58°F. (14°C.) are as comfortable as, and more healthful than, the old 72°F. (22°C.) setting ever was. People wear sweaters now in wintertime, and strangely enough, seem to have fewer respiratory ailments and colds, and even feel more alert and energetic.

- Turning off lights and appliances when you leave a room. Granted most appliances and lights generate heat, which, in turn, reduces the requirement on your heating system; but the low efficiency of a television set playing to an empty room is obviously wasteful.

- Drawing drapes or window shades when the sunlight is past, or on cloudy days. Later in this chapter we will discuss other window treatments that are even more effective in providing insulation.

By now you get the idea. Actually, we'll be discussing the conservation of energy in greater detail in chapter 5.

SOME FUNDAMENTAL CONSIDERATIONS

Now that you have exercised your intuition and common sense, let's take a more academic look at the subject of insulation. The basic problem is that most of us are gluttonous consumers of fossil-fuel energy. (U.S. citizens are, by far, the world's largest users of energy. We consume more than three times as much energy as the world average.) Currently, domestic energy consumption by residential users alone is approximately one-fifth of total U.S. energy consumption, or 6% of the total world demand.* This is how it breaks out:

U.S. ENERGY CONSUMPTION

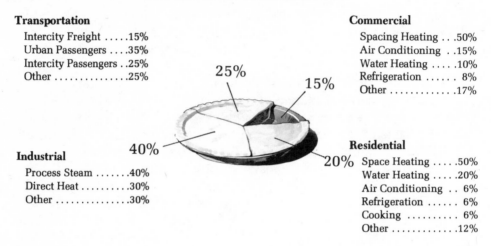

Transportation

Intercity Freight	.....15%
Urban Passengers	....35%
Intercity Passengers	..25%
Other	..............25%

Commercial

Spacing Heating	. . .50%
Air Conditioning	. .15%
Water Heating	.....10%
Refrigeration	...... 8%
Other	............17%

Industrial

Process Steam	.......40%
Direct Heat	..........30%
Other	..............30%

Residential

Space Heating	.....50%
Water Heating	.....20%
Air Conditioning	.. 6%
Refrigeration	...... 6%
Cooking	.... 6%
Other	............12%

Now that we've isolated this 20% residential slice of the energy pie, what can we do about it?

First, we must understand more about the fundamentals of heat energy. Of the five mechanisms (infiltration, conduction, convection, radiation, and evaporation) by which a house loses or gains heat, we will discuss the two, *infiltration* and *conduction*, which have the greatest impact on your home's heat-loss.

Infiltration occurs at any instant when there is a difference between the pressure of the air inside a house and that of the outside. This difference can be caused by wind, operation of a furnace, by a fire in the fireplace, or simply by temperature variations in the building. The result is always the same, though; an exchange of air occurs between the inside of your house and the outside. In winter, air that you paid hard dollars to heat is lost, and the icy air brought in must be heated if you are to be comfortable. Similar waste occurs in the summer if your home has air conditioning.

*NTIS (National Technical Information Service) publication PB 241–919, p. 2.

Infiltration can never be stopped. Some people think that a tightly sealed, well-insulated house doesn't allow enough air circulation to maintain a healthful environment. (Do you remember being taught in school that you must always sleep with the window open, no matter how cold or windy the outside conditions? It's been a popular belief for decades.) In most houses, 100% of the inside air is lost through cracks and seams, and replaced by outside air, *every hour!* Actually, it is now believed that only a 20%/hour air change is needed for normal ventilation,* and we may eventually find that standard excessive, too. Even with the installation of every possible insulation and conservation technique, it is probable that you could only limit infiltration to about 40%/hour, so don't feel restrained in sealing every air leak you can find. If you can't stop infiltration, how can you at least control it? Simple . . . through the use of storm doors and windows, weather stripping, and caulking any other openings you can find in your building. Another effective method is that of using north and northwest windbreaks made of evergreen trees and shrubs.

What about *conduction?* The moment there is a difference between the temperatures inside and outside your home, heat begins to flow through all exterior surfaces of the house, *always from the warm side toward the cold side.* The greater the difference in temperature (between inside and outside) the faster the heat flows! You can quickly visualize the impact of this on your heating and cooling needs during winter and summer seasons.

The answer to heat-flow resulting from conduction is *insulation.*

Before we launch our "how-to" portion, let's look more closely at this concept of heat-flow. Heat (energy) travels through different materials at different rates. The resistance to this travel is commonly referred to as the "R" value of the material. For example:

> 1. An 8-inch brick wall with 1-inch furring space, using flexible insulation and an inside surface of gypsum board, has an "R" value of 5
> 2. A wooden (5/16-inch plywood) wall with 2-inch batt insulation between studs, and an inside surface of gypsum board has an "R" value of 10

Thus, the frame house wall has *twice* the resistance to heat-flow as the brick house wall. Some people may find this surprising; we learned

*NTIS publication COM–75–11029, p. 16.

(in our "Three Little Pigs" story days) that brick homes were more secure than wood. However, they *are* heat-losers in the winter, and heat-gainers in the summer. We will be discussing the "R" values of various types of insulation shortly, but for now just remember, the higher the "R" rating, the better the resistance to heat-flow. A high "R" is *good* and a low value is *bad* when thinking in terms of building insulation!

Just a few more definitions relating to heat-flow:

We refer to heat energy in Btu's (British thermal units) when discussing the amount of energy that will travel through a particular material (or combination of materials) in a given period of time. A Btu is the amount of heat energy required to raise the temperature of 1 pound of water 1°F. (.56°C).

An average house will lose about 40 Btu's/square foot of heated area/hour, while a well-insulated house will lose only 20 to 25 Btu's/square foot/hour. This implies that a well-insulated house can cut heat-loss (from conduction) in half! What a marvelous opportunity for savings—both in fossil fuels and in dollars. *This explains why dollars invested in insulation offer a higher rate of return than any bank savings account* . . . why we've titled this chapter, "Your Best-Spent Dollars—Insulation."

HOW TO INSULATE

The most obvious methods of home insulation are adding storm doors and windows, installing weather stripping, and caulking around window and door frames. Others include attic floor insulation, exterior wall insulation, insulation of floors, etc. If someone were to ask us which of these insulating efforts gives the greatest return for effort and expense, we would list the top three priorities in this order:

1. above-ceiling insulation (attic)
2. window and door treatment
3. weather stripping and caulking

Beyond these priority efforts, the return on your dollar and time investment begins to diminish. This does not imply that you should not continue the insulation of crawl space or basement ductwork, walls, floors, etc. It means only that your initial effort should be where it will do the most in cutting your home's heat-loss.

If you wish to have your insulation installed by a contractor, the general advice offered in chapters 7 through 9 applies and will be helpful in carrying out your program. The balance of this chapter is designed for the do-it-yourselfer, and will stress that aspect of the job.

Kinds of insulation.

Materials

There is a wide variety of insulation materials, suitable for every conceivable situation, and then some. Your local building supply dealer will undoubtedly stock most of them, and should be an excellent source of information, with pamphlets and instruction sheets relating to their installation. (You might consider neighborhood cooperative insulation purchases, or off-season buying . . . for price advantages.) These are the general classifications of insulation:

Batts or rolls
- glass fiber
- rock wool

Loose fill
- glass fiber
- rock wool
- cellulose fiber (double-check its fire rating!)
- vermiculite

Rigid boards
- beadboard (expanded polystyrene)
- foamboard (extruded polystyrene)
- urethane
- glass fiber

Foam-in-place
- urea-formaldehyde (check for fire safety!)

The next question is "Where should each of these types of insulation be used, and why?" Information found in the Appendix will help answer this question, but for the most part, common sense must prevail. For instance, what type of material should be used in an attic?

Depending on the attic construction of your home, either batts, rolls, or loose-fill insulation might be most appropriate. If, for example, you have a small access panel leading to your unfinished attic area, short batts of insulating material would be the easiest to carry up into the work area, but rolls would do a better job, while being only slightly more difficult to handle. If your attic area has a stairway and a partial (or complete) floor, the job will be more difficult. In this case you must fill the cavity beneath the attic floor with insulation. This can be done with loose fill, or by stuffing batt or roll material into the cavity, or in the worst situation, by removing the floor, installing the insulation, and re-laying the floor.

Another word about the "R" ratings of insulation. We know that the higher the rating, the more effective the job that will be done. An "R" rating of 30 is not too high for any attic in the United States. This means you should have the equivalent of 10 to 12 inches of insulation of fiber glass above your ceilings. If your building supply house cannot supply you with the equivalent "R" ratings of various insulation materials, information available at the end of this chapter will answer your questions. Also, most insulating materials are conspicuously marked with their "R" rating.

Vapor Barriers

Another important factor with which you should be familiar before starting the job is *moisture control.* On many a cold day you have probably observed the moisture which condenses on windows. At times so much condenses that small rivulets run down the window and form puddles on the sill. When the weather is extreme, you may find this condensed water forming a coating of ice inside your window (particularly if the window is shielded from the room's warmth by drapes or shades). Moisture exists inside *all* homes. It comes from the lungs and bodies of persons living in the house, from plants, from cooking, from showering and laundry, and, in some cases, from humidifiers that provide a comfortable level of humidity.

To prevent this humidity from also condensing in your home's insulation (with the many attendant problems that could occur from water or ice accumulating in your walls and ceilings), a "vapor barrier" must be used along with all porous-type insulation. This is the water-proof paper or foil which covers one side of some (batt or roll) insulating material. The barrier discourages water vapor (air moisture that we have just described condensing on your cold windows) from passing through, or condensing in the insulation. *The vapor barrier should always be nearest the inside of the house* (facing in, to the "lived-in" [warm] portion of your home) unless competent specialists advise you otherwise for problem areas.

When adding more insulation to an already insulated attic floor, *do not* use an additional vapor barrier. If the additional insulation you plan to use is available only with a vapor barrier, you must either remove it, or perforate the barrier with a knife, so that air can move through it freely. Again, we restate that if you are installing insulation where none presently exists, you must have a vapor barrier, and it should face the lived-in area of the home where the humidity originates. Incidentally, many paint manufacturers have recently been producing vapor-proof paint, which permits you to paint your walls and install a vapor barrier at the stroke of a brush or roller.

Planning the Job

So, let's start planning your first-priority task, insulating your attic. We'll presume that it is an unfinished attic with an access panel opening. The first action on your part is to put a ladder up the access panel, and with a good flashlight, take a look around. Once you're up there,

The gypsum board ceiling of the room below is visible.

This is what the job should look like after you've finished.

we'll further assume you find an uneven covering of loose insulation such as rock wool. In some areas of the attic it covers the rafters (about 6 inches deep), while in other areas the gypsum board ceiling of the room below is visible. You must also check the rafter spacing. Sixteen inches is standard. With this in mind, close the access opening, descend the ladder, and begin your planning.

Your first consideration is insulation material. The addition of 6-inch-thick, 15-inch-wide rolls of fiberglass insulation (R-19) will add enough to your existing insulation to do the job, assuming your present insulation averages 4 inches deep with an R-11 value. (The table at the end of this chapter will give you "R" values.) We've presumed the use of fiberglass because it's readily available, inexpensive, easy to work with, and generally provides a consistent, effective job, even when in-stalled by a novice. Calculate the area to be covered (you can subtract 10% for the space taken by the rafters themselves) and your building supply house will tell you how many rolls you will need. You might even call two or three suppliers, to get the most competitive price. Make sure the insulation you purchase has no vapor barrier, since you will be adding it to existing insulation.

Next, set a specific time to do the job. A normal unfinished attic of approximately 1,000 square feet should take no more than two or three days. Plan a three-day weekend, and make a list of necessary tools.

You will need the following, so have them and the insulating material ready, before your three-day weekend begins.

• A sharp knife or large shears, with which to cut the rolls (or batts) of insulation.

• A measuring tape, or folding (carpenter's) ruler.

• A child's hockey stick. We find this the most useful tool for positioning the insulation between the rafters near the eaves, where a person's reach is restricted because of the roof slope.

• Walkboards. Cut several so that they span three to four rafters. Place a small lip on the ends, so that the ends of the walkboard "lock" over the edge of the rafter.
(Nothing is quite so surprising as stepping on the end of a poorly positioned walkboard, overhanging a rafter, and having it pivot downward as your foot crashes through the ceiling below.)

• Protective gloves to wear when handling the insulation.

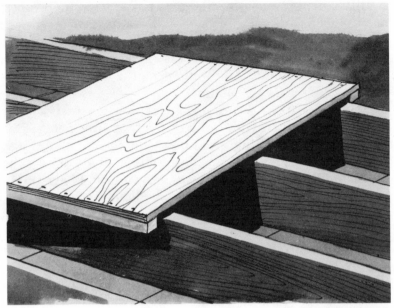

Rafter walkboards.

• A surgical mask (available at any drugstore for under 50¢). Because most attics have accumulated years of dust, we find it very helpful to cover the nose and mouth.

• An electric trouble light and extension cord.

• A hammer and nails, so you can make a place to hang your trouble light.

• A staple gun may come in handy when you insulate the backside of your access panel, and in occasionally tacking the insulation to the rafters.

• Another good idea is to have a large glass or bottle of cool water, sealed with plastic wrap, with a straw inserted. That will permit you to sip when you're thirsty, but will keep out the dust.

• A portable radio will also help pass the time with some good background music.

With your tools and insulation materials available, you are now ready to start. Dress properly for the job. Cover as much of your body as possible: long sleeves, dungarees, and perhaps a head covering (and a sweatband for your brow). It is a dirty, but important job that you are about to start.

The "compleat" insulator.

If you've chosen rolls of insulation material, they come packed four rolls to a bale. Don't be surprised when you open your first bale. The insulation is tightly compressed when it is packaged, and quickly begins to expand when the bale is broken open. It appears alive, and grows to about four times its former volume in a few moments. In fact, it is wise to take the unopened bale through your attic access opening (if it fits) rather than make four trips up and down the ladder with each expanded roll. It is perfectly safe to break open the bales in the attic area.

Four rolls come compressed in one bale. Expanded roll next to unopened bale.

Above Ceiling (Attic)

Start your attic insulation job by establishing a "base of operations." By this we mean a safe, comfortable spot near the access opening where your tools and materials can be located. Run an extension cord and hang your trouble light in the most convenient spot.

Start working at the farthest corner (*never* start in the center and work toward the sides); systematically take your tools and material to that end of the attic. Be particularly careful not to put a foot on the ceiling material down between the rafters; you can easily go through. Use the walkboards you've prepared. Also, watch out for roofing nails protruding through the roof above your head. *Be careful.*

Another precaution is to treat electrical wiring with care. Don't try to pull it, or bend it out of the way. Work around it, or over and under. Also, don't cover the eaves and gable vents with insulation. Remember, you are only insulating over the living area of your home. A free flow of air must be available from the eaves and gable ventilators (to prevent the accumulation of moisture).

You can easily go through.

Watch out for protruding nails.

Treat electrical wiring with care.

Don't cover the gable vents.

Light-fixture housings on the floor of the attic (from recessed fixtures in the room below) should be allowed "breathing room" when you insulate. Do not cover them or pack your insulation tightly around them. The lights they contain can give off enough heat to be dangerous if you do not allow clearance for the heat to dissipate. (Unfortunately, this is lost heat energy in the winter, but in the summer it helps keep the room cooler than it might otherwise be.)

Allow breathing room around recessed fixtures.

If the roof pitch is low, the most difficult part of your job will be laying the insulation in the far reaches, where the roof rafters join the attic floor. The reason for the difficulty will be obvious once you're in the attic, doing the job. You just can't reach into that narrow space between the attic floor and roof. And that is precisely the reason we mentioned a

Use the working end.

child's hockey stick in our list of recommended tools. Use the working end (not the handle) of the hockey stick to push the insulation into position as you lay it between the rafters. Rather than attempting to work with a full or nearly full roll of insulation, cut off about an 8- to 10-foot length. Position the cutoff end about the same (8 to 10 feet) distance from the eave, and using the hockey stick, work the insulation between the rafters, toward the eaves (roof/floor juncture). Remember, if your roof overhangs the building wall, insulate only over the ceiling of the room below; do not stuff the insulation all the way to the corner. You must allow for air circulation in the overhang portion.

Once you've repeated this process between each of the rafters near the low-slope edges of your attic, the worst is over. Complete the job by filling all spaces between the rafters with your insulation material, working your way toward the access opening.

Don't ignore the small finishing touches. Take the time to fill small openings and narrow spaces, and staple insulation to the access opening panel, etc. Once you've gone this far, it would be a shame not to put the final touches on the job so that you can enjoy the confidence of knowing that you've controlled the area of greatest heat-loss from your home—the attic!

Window and Door Treatment

The most obvious solution one thinks of regarding heat-loss from windows and doors is the addition of storm windows and doors. This is

Indoor insulating shutters.

appropriate, and should be given prime attention, but don't stop there. More can be done, frequently with very inexpensive techniques, to minimize your building's heat-loss. Let's start by discussing windows, since almost all of the openings in the shell of your home are windows. The illustrations shown indicate a solution to heat-loss which a family in the Northeast conceived: simple, attractive, indoor window shutters. Because the home has casement-type windows rather than conventional double-hung ones, the family had difficulty in obtaining storm windows. The solution to this problem was an idea that turned out to be very inexpensive and extremely effective.

The basic construction of their shutters is very simple. The material used was common flakeboard, 3/4 inch thick, cut to fit the inside of the window frame. Although the shutters would be somewhat more effec-

Insulating shutters on a bathroom window, open at right and closed at left.

tive if made from an insulating material such as urethane covered with plaster, the increment of improvement is small compared to the increased difficulty of construction. Small hinges, simple pull knobs, and magnetic door latches were the other components. A narrow molding strip served as the bottom shutter striking surface, which did two things. It provided a stop for each shutter panel, assuring that it closed on the same plane and fitted evenly, and it prevented air leakage at the bottom of the shutter where the cold air would have tended to sneak under anything but perfect-fitting shutter panels. The magnetic latches acted as the striking surfaces at the top of the window, providing the upper shutter alignment, as well as the latching function.

The results enjoyed by the family were dramatic, particularly where they constructed folding shutters for use on the large window. The room, which was almost useless in winter months because of cold drafts from the prevailing northwest winds, is now draftless, snug, and is used with great pleasure by all members of the family.

Window shades and drapes can also play an important role in minimizing your home's heat-loss particularly if they touch the sill or

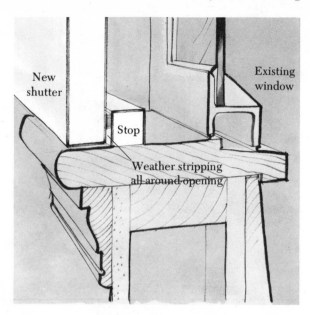

New shutter

Existing window

Stop

Weather stripping all around opening

A narrow molding strip serves as the bottom shutter striking surface.

floor, creating a draft-seal. Windows exposed to sunshine on winter days should have drapes and shades open, to allow the solar energy to warm the room. When the sunlight is gone, the shades and drapes should be drawn to prevent the excessive heat-loss which would otherwise occur through the windows. Just the opposite should be done with the penetrating heat of the summer sun. Draw the shades and drapes to help cool areas that would otherwise be overheated (and in turn cause your ventilating or air-conditioning system to work more than necessary) by the summer sun. Again, our advice is nothing more than common sense, observation, and awareness; but it can make a difference in your energy costs and in your living comfort.

Just in case you haven't heard about Beadwall, a few words about it may be appropriate. It's a window-type unit consisting of two layers of glass, or other glazing material, between which an insulating mass of styrene pellets can be blown—and later removed—simply by flicking a switch. The switch operates a small blower which moves the pellets. Somewhere nearby, of course, a container to accommodate the evacuated pellets must be provided. Zomeworks, of Albuquerque, is the manufacturer, and we hear good reports about this product, but we do wonder what verb to use in describing the actions of a Beadwall. Does one say, "Please fill the Beadwall, Fred," or "Close the wall," or what? If not soon resolved, this could escalate to become a serious problem in interpersonal relationships.

Typical Beadwall when "open" admits solar energy through its double panes.

Typical Beadwall when partly "closed," forming an insulated wall between warm room and cold night. A flick of the switch removes the insulating pellets between the panes.

One clever solar manufacturer produces venetian blinds that are energy absorbing (dark) on one side, and light reflecting (aluminized) on the other. On winter days they are drawn so that the dark side absorbs the sun's energy. Convection currents between the slats allow this warmth to circulate freely into the room, with cooler air being drawn up into the slats. On summer days the slats are reversed, so that the sunlight is reflected back to a great extent, thus helping to cool the room.

One of the most difficult "windows" to insulate is the skylight. Because they've been used rather infrequently in most homes in the past, we don't often think of them. But as the energy shortage worsens and construction becomes more conservation-oriented, we *will* be seeing more of them. As glass window areas are reduced in houses, more designers will turn to the strategic use of skylights in home concepts. And as the acceptance of earth-sheltered housing grows, skylights will have an even more important role to play. A simple method of insulating skylights is described at the end of this chapter.

By now it is evident that awareness and a willingness to act on what your common sense tells you can play a major role in reducing heat-loss through windows; but what about doors? The most effective door treatment (in addition to the obvious storm door) is the "air-lock" vestibule. Many homes lend themselves to the addition of an air-lock simply by

An air-lock vestibule.

adding a small partition and door. This is frequently the case where a small closet borders your front entrance or foyer area. An air-lock is especially effective in "active" households where young children are constantly in and out all day long.

Weather Stripping and Caulking

This is the item mentioned earlier in the chapter as having the third priority in terms of investment value (giving the greatest return for your dollars and effort) against your home's heat-loss. Start out by assessing your building's needs. This is best done on a windy winter day, when the needs will be blatantly obvious. Arm yourself with paper and pencil for note-taking so items won't be forgotten during your assessment. Check each window and frame carefully for drafts, and for loose or miss-

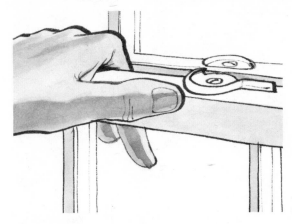

At times large spaces exist between window frames.

ing glazing material (putty). At times, large spaces (large enough for a pencil to penetrate) exist between window frames and walls; an obvious need for weather stripping. Check exterior doors and their frames.

A walk-around and visual check on the outside are also useful. Pay particular attention to the caulking around window and door frames. If it is hard, dry, and cracked, it should be removed and replaced. Also, check each of the corners on your building carefully. Caulking is often needed where the walls of your building come together, or meet the roof. Another area to check is under the lowest line of shingles or clapboard on frame construction houses. Pay attention to other minor openings such as those at exhaust fans, clothes dryer exhaust hoods, through-the-wall water faucet outlets, electrical power lines, etc.

Once you have completed your survey, prepare for the job with a caulking gun and tubes of caulking. These are available at your local building supply house, where you can get instructions if necessary. Several types of weather stripping are also available. Felt fiber strips can easily be tacked around the inside of door and window frames, and even more convenient is adhesive-backed foam rubber stripping. Be sure to discuss your needs with your local building supply store manager. He

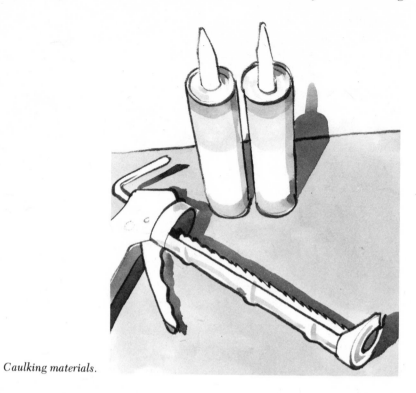

Caulking materials.

Opening around water faucet pipe.

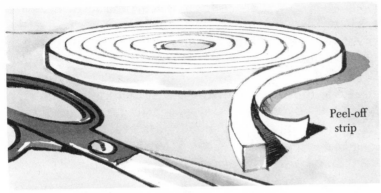

Adhesive-backed foam rubber stripping.

will guide you to the various materials available for controlling your heat-leaks.

Basement and Crawl Space

If you have covered the top-priority insulation tasks (attic, windows and doors, weather stripping and caulking) and are planning to continue in your basement or crawl space, an excellent guide to follow, with easy-to-understand illustrations, is a government publication called *In the Bank* . . . or *Up the Chimney?*

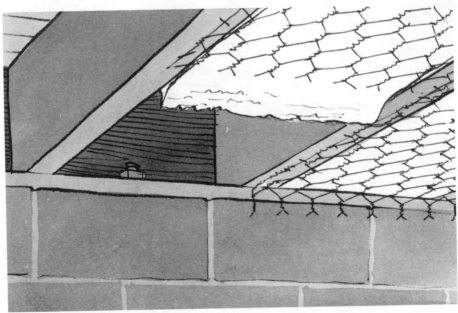

Chicken-wire retainers.

If your basement or crawl space is unheated, the most effective treatment for it is to insulate the underside of the floor above. Generally, the overhead joists will be on standard 16-inch centers with occasional cross bracing. One easy method is to use fiberglass batts or rolls. Insulation can be added to the space between joists, and retained there by tacking chicken wire to the joists (allowing the insulation to rest on the chicken wire) or merely by crisscrossing string under the insulation once it is positioned. The vapor barrier should be facing the room above; an aluminized paper barrier would be effective in this application. Cut the length of insulation so that it can be easily installed between the joist cross braces.

Exposed ductwork and pipe should also be insulated. The photos here show a typical ductwork installation before and after being insulated. Ductwork is best insulated using aluminum-foil-backed insulat-

*Typical crawl-space duct-
work before insulation is installed.*

*This is how the crawl-space
ductwork appears after being insulated.*

ing material, 2 inches thick, designed for duct needs. In this case, the foil (vapor barrier) must be on the outside, with the insulating material placed against the ductwork. Wide duct tape is available for fastening the insulation material to the ductwork. Using the duct tape, you should first seal any leaks evident at the corners or bends of the ductwork. Then, starting at one end of the exposed ductwork, apply the insulation and *seal it carefully* with the duct tape. If your ducts are used for central air conditioning as well as heating, it is *particularly important* that the insulation be thoroughly and completely sealed. This will prevent condensation from forming between the duct and the insulation during the summer cooling season.

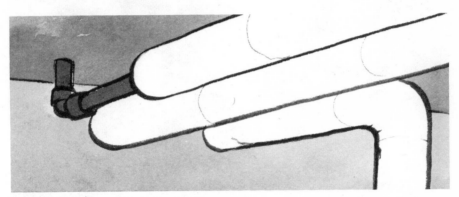

Pipe insulation partially completed.

Water pipes (both hot and cold) should also be insulated. Various types of pipe insulating materials can be purchased at your building supply outlet. Some are suited for use in tight quarters (where your access to the pipes is limited), while others are more effective for long, open runs of exposed pipe.

If your basement is heated, its treatment will be quite different. In this case there is no need to insulate the underside of the floor above or the exposed ductwork. The warmth (or cooling) lost from the ductwork will benefit the basement living area, and will warm (or cool) the underside of the floor above. On the other hand, hot and cold water pipes should be insulated.

Assuming your basement outer walls are cinder block construction and dry, they can be insulated in either of two ways. The first is to completely frame the inside of the basement using standard 2x4 lumber, and then insulate between the studs as you would any other frame construction. After the insulation is installed (with the vapor barrier facing into

Insulate between new studs.

*Rigidboard insulation
placed between new furring strips.*

the room), the construction can be finished with either paneling or 1/2-inch gypsum wallboard.

The alternate method is to nail furring strips to the inside surface of the basement's outer walls (24-inch spacing between the strips) and place panels of insulating rigidboard between the furring strips (it comes in 24- or 48-inch widths). The rigidboard is generally made of polystyrene or urethane, (with an "R" rating of about 4 per inch) but because of fire hazard (they may emit toxic gas when burned) must be covered with 1/2-inch gypsum wall board. It is recommended that rigidboard be installed by a qualified contractor, unless you've had previous experience working with this material.

The poured concrete floor of your basement will also be the source of some heat-loss, but the amount will vary depending on how deep the basement is in the earth. The deeper in the ground, the less heat-loss that will occur. This will be discussed in more detail later in the chapter when we address the subject of poured concrete floors.

Don't forget that your crawl space (if you have one) should be vented to the outside. This prevents a buildup of moisture under your home, but at the same time it mandates that your under-floor construction should be well insulated, as should all ducts and pipes that run through the crawl space.

Furnace and Hot Water Heater

Finally, there are two more potential sources of heat-loss in your basement area which need your attention. They are your heating system furnace and your hot water heating system. Both will benefit from insulation, but special care must be exercised because of chimney flue heat. We will explain that in just a moment, but first, let's talk about the furnace. Presuming your system is forced hot air, your furnace has three basic sections; the air intake (which generally houses the blower), the firebox, and the hot-air plenum, which delivers the warm air to your

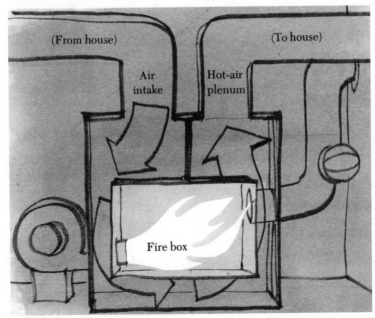

Your furnace has three basic sections.

system of ductwork. The warm-air plenum also houses the cooling coils of your air-conditioning system (if you have built-in air conditioning). It is this hot-air plenum (the upper portion of your furnace) that can benefit from insulation. You can use the same 2-inch aluminum-foil-backed duct material that we discussed for use on the basement/crawl space ductwork. Again, use your wide duct tape to first seal any air leaks; then apply the insulating material (foil facing outward) and seal it in place with the duct tape. Be sure it is completely sealed so that condensation will not form between the plenum and the insulation during the summer air-conditioning season. Now, about that "special care" mentioned earlier in this paragraph: When insulating your furnace plenum and ductwork, you will notice a flue pipe coming from the furnace fire box, and going to your chimney. *Do not* cover this flue pipe with insulation since the flue can get quite hot in winter months when the furnace is running—more than you'd like it to be—and could possibly cause a fire (if it were wrapped with paper-backed insulation).

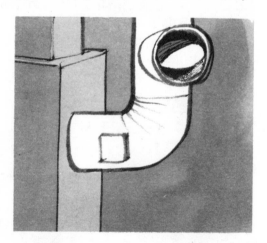

Furnace flue pipe. Do not insulate!

Your hot water tank is the final basement item which should be insulated. Manufacturers of insulation, such as John Manville, offer water heater insulation kits, similar in concept to the tea cozy your mother may have used to keep the teapot hot. For electric hot water heaters, the insulation kit is nothing more than a tube (closed on the top) which you slip over the water storage tank. The gas-fired hot water storage tank uses a similar tube with an opening for the chimney flue. (The same word of caution applies here as with your furnace flue; it gets quite hot at times, so avoid intimate contact between the flue and any paper-backed insulation.)

It is not necessary to use a manufacturer's insulation kit; you can accomplish the same result with your roll or batt insulation, held in place and sealed with duct tape. One important fact to keep in mind is that of all the insulating tasks to be accomplished in your basement/crawl space area, the insulation of your hot water storage tank is probably the most important. This effort will repay you for the material and time invested more quickly than other basement efforts.

Wall-Insulating Techniques

Carefully remove the baseboard and peek behind it.

If your home was built before the practice of wall insulation became widespread, there's a good chance that your walls contain no insulation. If you're not sure, but suspect that to be the case, this is what you should do: Choose a wall that you think has been losing heat and allowing crisp winter breezes to enter. Carefully remove the molding from the base of the wall. (Remove one panel carefully if it is a paneled room.) With the molding removed, you should be able to see beneath the gypsum wallboard, but if not, cut away a small portion, less than the height of the molding (so the cut won't show after the molding is replaced). Now you can determine whether or not your walls are insulated.

Another possible way of checking is to remove switch plate covers or wall electrical outlet covers. With the help of a flashlight you may be able to peek behind the electrical fixture box to determine if insulation exists in your wall. The upper portions of some walls which contain loose rock wool insulation may appear to be drafty. This is not necessarily illogical, since the rock wool often "settles" during the years after being installed, creating a space for additional material, which would improve the insulation value of the wall.

Peek behind your electrical outlet cover . . .

. . . with the help of a flashlight.

Assuming the worst, that your walls are uninsulated, what can be done? Actually, there is little the do-it-yourselfer can do, short of tearing down the inside walls, insulating, and refinishing the room with new walls. But *there is* a solution if you call in a professional insulation contractor. He can insulate your walls by installing glass wool or rock wool in the following manner. He will generally work from the outside, and will start by removing the line of shingles or clapboard that is outside the upper portion of your room wall, near the ceiling. After locating the vertical studs, he will drill a hole in the undersheathing between each of the wall studs. Into this hole he will insert a hose through which the insulation is blown. In this manner, each cavity (formed by the wall studs) will be filled with insulating material. When finished, he will seal each of the access holes he drilled earlier, and will replace the clapboard or shingles, leaving the wall looking unchanged, but actually much better insulated than before. In the same way, "settled" rock-wool insulation can be topped-off in any problem wall. Because of the

Loose insulation blown into wall cavity.

expense involved, this course of action should be saved for serious problem situations.

Another material used by contractors is plastic foam (urea formaldehyde), installed under pressure. It quickly hardens to form solid insulation. Because the materials and the technique are relatively new, it is important that you use the services of a *qualified* professional. With this foam-in material, you have no vapor barrier problems (since moisture cannot penetrate the foam), but with blown-in rock wool or other loose fill (like cellulose) it is important that your walls be painted with a vapor barrier paint, or that you wallpaper with a vapor-proof, vinyl-base wallpaper.

Poured Concrete Floors

Following World War II, the need for single-family dwellings gave rise to the construction of many massive housing developments. One technique used to expedite the construction process was the elimination of basements, and the use of mass-produced and prefabricated components. A favorite scheme used in many developments of this vintage was to pour concrete slabs, upon which the dwellings were then erected. The idea caught on quickly because of standardized construction techniques which generated lower cost and speedy construction. In many developments, various enhancements were added to the poured concrete pad, such as hot water pipes embedded in the concrete to form a radiant heating system for the home. In the time of energy-plenty, this method of heating looked enticing. The concrete floors (never quite "first class") were upgraded by being warm and comfortable to walk upon (even barefoot) during cold winter days. The radiated heat would then rise comfortably to heat the room. At least it did until the owners began cushioning the hard concrete floors with carpet padding, covered in turn with thick, plush carpets . . . *very effective insulation* where it was not wanted! But that didn't matter since the owner merely pushed up the thermostat to compensate for the insulating effect of the pad and carpet . . . until the dawn of realization fell upon these homeowners that the age of cheap energy was gone. Suddenly the question was asked, "Why are these concrete slabs such heat-drains?" You might think that once the earth under them had warmed up, it would behave as insulation, and the heat would be delivered only to the home, not the already-warm earth below. The problem was that this theory did not go far enough. The earth *beneath* the concrete pad was considered, but not the earth *surrounding the ends of the pad.*

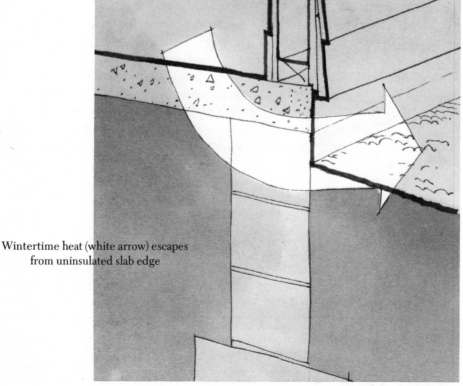

Wintertime heat (white arrow) escapes
from uninsulated slab edge

Slab-edge heat loss.

In order to better understand why poured concrete pads are heat-drains, and what can be done to improve this situation, look at the sketches here. The earth at and near the ground surface is almost as warm or cold as the air that it touches. As you probe deeper into the earth, the temperature becomes more stable. In any geographic location, the frost never reaches below a certain depth, because of the insulating quality of the earth. So we see that the deeper we go into the earth, the more stable the temperature; and conversely, the closer to the surface, the less stable the temperature. The result of this fact (which was not fully appreciated by most architects and builders until recently and is still not appreciated by some) was that the heat from the concrete slab literally poured *from its edges* into the air and cold earth near the surface.

Once this heat-loss mechanism is understood, the technique for correcting the situation becomes relatively easy: insulate the edges of the poured concrete pad. This is most easily done by digging a trench around the periphery of the pad and installing rigidboard insulation.

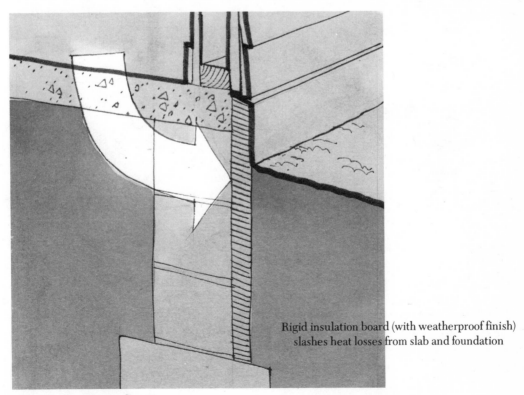

Rigid insulation board (with weatherproof finish) slashes heat losses from slab and foundation

Slab-edge heat loss corrected.

The accepted method is to apply waterproofing to the exposed ends of the concrete and then apply the insulation and backfill. The previous sketch provides the detail.

Insulating your poured concrete slab also benefits you in the hot weather. You'll recall that heat always flows from hot to cold. In the summer, the hot summer air and surface earth in contact with the pad edges, transmits its heat to the massive slab. The slab in turn radiates the heat into your rooms, making them less comfortable, and giving your air conditioner an extra burden. This heat-gain situation will also benefit from edge insulation.

INSULATION AND SOLAR ENERGY

We can't stress often enough that the key to lower fuel usage is *insulation* first, and consideration of other techniques such as solar heating second. Assuming that you are now well aware of this fact, and will improve your home heating efficiency through the use of insulation, the final paragraphs of this chapter discuss special insulation considerations which relate to solar components.

Collectors

Whether your solar heating system is an air type or uses liquid (water, or water and antifreeze), proper insulation will increase its operating efficiency. You may recall that in chapter 2, where we discussed solar flat-plate collectors, the absorber plate was shown to have insulation between it and the underside of the collector. Most of you will not be building your own collectors, so your concern with solar collector insulation will be limited to a careful review of various manufacturers' specifications. Be aware of the "R" rating of the insulation behind each type of collector, since manufacturers' statements can be misleading. For instance, two manufacturers may claim 2 inches of insulation behind their absorber plate. The first might be glass wool, with an "R" rating of 6, while the other may be rigidboard—with an "R" rating of 9 (a 50% greater resistance to heat-loss).

One comment we received from a solar homeowner states, "In the future I plan on using more aluminum foil on the roof as a reflector and more insulation in the [flat-plate] collectors."

So you can see, the insulation characteristics of the collector panels are important, and should be given consideration *before* you buy.

Ducts and Pipes

The conduits which carry solar warmth from your flat-plate collectors to your energy storage area also need complete insulation. This is true whether your solar heating system is an air type, using large air ducts with rock-bin energy storage, or a fluid type system, using conventional pipes and fittings leading to a water storage tank. Insulation techniques described earlier in the chapter should be employed on all ducts and pipes.

This was amplified by another solar homeowner who wrote, "My ductwork was insulated where it was outside [the shell of the house], but should also have been insulated on the interior [through the garage, attic, basement, etc.].

The only exception might be where the ducts and pipes run through the living areas of your home. Some designers deliberately leave these portions of the ductwork uninsulated. Their objective is to allow some heat-loss from the ducts or pipes to occur in the heated portion of the home to help warm the living area. Naturally, pipes and ducts can be unsightly in the "living" portions of your home if they are not handled in an esthetically pleasing way. Most people, though, prefer to route them through "nonliving" portions of their homes (attic, walls, basement, garage, crawl space, closet, etc.), and there, adequate insulation is super-important.

Heat Storage

The more common methods now used for storing solar energy during sunless periods are rock-bins for air systems and water tanks for fluid systems. Proper insulation of these energy storage devices is quite critical. Earlier in this chapter we noted that the insulation of your (domestic) hot water storage tank is probably the most important task to be accomplished when insulating your basement/crawl space. This applies equally to your heat storage areas. After going to the expense and effort of having a solar system installed, it would be foolish to allow the collected (and stored) energy to leak away into the surrounding soil or basement; yet many existing installations suffer from this problem. Often, too much design attention was put on collector and storage size, location, etc., and not on adequate insulation. Don't make this mistake on your installation!

Because there are so many possibilities for liquid heat storage (steel tanks, fiberglass tanks, vinyl fabric tanks, concrete tanks; above, partially in, under earth) and an equal number of possibilities for air

indoor
hinged
panels

indoor
bifold
panels

indoor
sliding
panel

indoor
overhead
panel

indoor
skylight
panel

Dramatically Reduce Window and Door Heat Losses

A large proportion of the wintertime heat-loss from most buildings occurs through and around windows, doors, and skylights. Whether they're of single or double glass, whether they're weatherstripped or not, a surprising amount of precious heat escapes in that way.

If you can cover those openings with insulation, particularly during winter nights but also, when possible, during those daylight hours when no direct sunlight is entering the room, you'll save a lot of money.

We and our customers have gotten some impressive results from these simple devices. They can be made of opaque or translucent insulation board, or even 3/4-inch plywood or chipboard, with or without wood frames. Ordinary hinges, magnetic catches, and grooved tracks or guides complete the ensemble, with sponge or felt weather-stripping being a further refinement.

Any kind of finish is acceptable as long as moisture cannot easily get through to the cold surfaces, and the panels fit snugly, particularly at the side and bottom edges. (The column of icy air trapped behind the panels, being heavier than room air, will try to pour from all side and bottom joints.) The panels, even when only partly closed, can reduce summertime heat-gain, too, and they make good privacy devices.

heat storage—plus all the combinations of rock storage and water tanks—it is impossible for us to suggest a "best" or "optimum" insulation technique for you to use. As a general rule, an "R" value of at least 12 should be provided for your entire energy storage area. Materials (or combinations of insulating material) providing this "R" value will assure you that your collected and stored energy will be waiting to serve you on those sunless days.

We have written this chapter for the owners of existing homes, but all the ideas and principles mentioned apply with equal force to the construction of new homes, where there is no excuse for doing anything but the best kind of insulating work. Do it well. As we said, it's your best investment.

APPENDIX/CHAPTER 3

WHAT THE R VALUE OF THE INSULATION SHOULD BE*

UNFINISHED ATTIC, NO FLOOR

Batts, blankets, or loose fill in the floor between the joists:

THICKNESS OF EXISTING INSULATION	HOW MUCH TO ADD	HOW MUCH TO ADD IF YOU HAVE ELECTRIC HEAT OR IF YOU HAVE OIL HEAT AND LIVE IN A COLD CLIMATE	HOW MUCH TO ADD IF YOU HAVE ELECTRIC HEAT AND LIVE IN A COLD CLIMATE
0"	R-22	R-30	R-38
0"-2"	R-11	R-22	R-30
2"-4"	R-11	R-19	R-22
4"-6"	none	R-11	R-19

FINISHED ATTIC

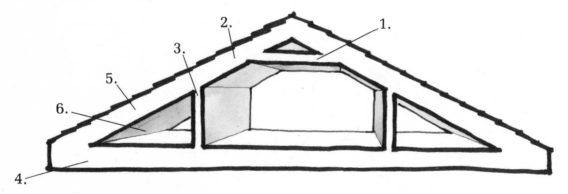

1. Attic Ceiling—See the table above under Unfinished Attic, No Floor.

2. Rafter — Contractor fills completely with blow-in insulation.

3. Knee Walls — Insulate (5), Outer Attic Rafters instead.

4. Outer Attic Floors — Insulate (5), Outer Attic Rafters instead.

5. Outer Attic Rafters — Add batts or blankets. If there is existing insulation in (3)

*Taken from *In the Bank . . . or Up the Chimney?*

and (4), add R-11. If there is no existing insulation in (3) and (4), add R-19.

6. End Walls — Add batts or blankets, R-11.

UNFINISHED ATTIC WITH FLOOR

A. *Do-it-yourself or Contractor Installed:*

Between the collar beams—Follow the guidelines above in Unfinished Attic, No Floor.

Rafters and end walls—Buy insulation thick enough to fill the space available (usually R-19 for the rafters and R-11 for the end walls).

B. *Contractor Installed*

Contractor blows loose-fill insulation under the floor. Fill this space completely.

Frame Walls—Contractor blows in insulation to fill the space inside the walls.

Crawl Space—R-11 batts or blankets against the wall and the edge of the floor.

Floors—R-11 batts or blankets between the floor joists, *foil faced*.

Basement Walls—R-7 batts or blankets between wall studs. Note: Use R-11 if R-7 is not available.

Kind of Insulation to Buy

BATTS—*glass fiber, rock wool*

Where they're used to insulate:
    unfinished attic floor
    unfinished attic rafters
    underside of floors

best suited for standard joist or rafter spacing of 16" or 24", and space between joists relatively free of obstructions

cut in sections 15 inches to 23 inches wide, 1" to 7" thick, 4' or 8' long

with or without a vapor barrier backing—if you need one and can't get it, buy polyethylene except that to be used to insulate the underside of floors

easy to handle because of relatively small size

use will result in more waste from trimming sections than use of blankets

fire resistant, moisture resistant

FOAMED IN PLACE—*urea-formaldehyde*

Where it's used to insulate:
    finished frame walls
    unfinished attic floor

may have higher insulating value than blown-in materials

more expensive than blown-in materials

quality of application to date has been very inconsistent — choose a qualified contractor who will guarantee his work

fire resistant, moisture resistant

BLANKETS—*glass fiber, rock wool*

Where they're used to insulate:
    unfinished attic floor
    unfinished attic rafters
    underside of floors

best suited for standard joist or rafter spacing of 16" or 24", and space between joists relatively free of obstructions

cut in sections 15" or 23" wide, 1" to 7" thick in rolls to be cut to length by the installer

with or without a vapor barrier backing

a little more difficult to handle than batts because of size

fire resistant, moisture resistant

RIGIDBOARD—*extruded polystyrene beadboard (expanded polystyrene) urethane board, glass fiber*

Where it's used to insulate:
    basement wall

NOTE: Polystyrene and urethane rigidboard insulation should only be installed by a contractor. They must be covered with 1/2" gypsum wallboard to assure fire safety.

extruded polystyrene and urethane are their own vapor barriers, beadboard and glass fiber are not

high insulating value for relatively small thicknesses, particularly urethane

comes in 24" or 48" widths

variety of thickness from 3/4" to 4"

LOOSE   FILL (poured-in)—*glass fiber, rock wool, cellulosic fiber, vermiculite, perlite*

Where it's used to insulate:
    unfinished attic floor

vapor barrier bought and applied separately

best suited for nonstandard or irregular joist spacing or when space between joists has many obstructions

glass fiber and rock wool are fire resistant and moisture resistant

cellulosic fiber chemically treated to be fire resistant and moisture resistant; treatment not yet proven to be heat resistant, may break down in a hot attic; check to be sure that bags indicate material meets Federal Specifications. If they do, they'll be clearly labeled.

cellulosic fiber has about 30% more insulation value than rock wool for the same installed thickness (this can be important in walls or under attic floors)

vermiculite is significantly more expensive but can be poured into smaller areas

vermiculite and perlite have about the same insulating value
all are easy to install

LOOSE FILL (blown-in)—*glass fiber, rock wool, cellulosic fiber*

Where it's used to insulate:
    unfinished attic floor
    finished attic floor
    finished frame walls
    underside of floors

vapor barrier bought separately

same physical properties as poured-in loose fill

because it consists of smaller tufts, cellulosic fiber gets into small nooks and corners more consistently than rock wool or glass fiber when blown into closed spaces such as walls or joist spaces

when any of these materials are blown into a closed space enough must be blown in to fill the whole space

How Thick Your Insulation Should Be
TYPE OF INSULATION

	BATTS OR BLANKETS		LOOSE FILL (POURED-IN)			
	glass fiber	rock wool	glass fiber	rock wool	cellulosic fiber	
R-11	3½"-4"	3"	5"	4"	3"	R-11
R-19	6"-6½"	5¼"	8"-9"	6"-7"	5"	R-19
R-22	6½"	6"	10"	7"-8"	6"	R-22
R-30	9½"-10½"*	9"*	13"-14"	10"-11"	8"	R-30
R-38	12"-13"*	10½"*	17"-18"	13"-14"	10"-11"	R-38

*two batts or blankets required

BIBLIOGRAPHY

Energy Primer. Whole Earth Truck Store and contributing authors. Portola Institute, Menlo Park, CA 94025.
A comprehensive book about renewable sources of energy. The focus is on small-scale systems which can be applied to the needs of the individual, small group, or community.

44 Ways to Build Energy Conservation into Your Homes. Owens/Corning Fiberglas Corp., Toledo, OH 43604. 1975. Pub. No. 5–BL–7055–A.
All ideas in this booklet are designed to conserve energy and save money in the home.

Home Energy Saver's Workbook. Federal Energy Administration, Nov. 1976. 29 pp., FEA/D–76/362. Available from the Superintendent of Documents, U.S. Government Printing Office, Washington, DC 20422; order number 041–018–00116–8. $.35.*
Helps reader determine what measure will make a home more energy-efficient and what the reader can expect to save in home heating and cooling costs by taking these measures.

Household Energy Game. Thomas W. Smith and John Jenkins. University of Wisconsin, Madison, WI 53706. December 1974. Available from National Technical Information Service, Springfield, Va 22161; order number COM–75–10304.

How to Save Money by Insulating Your Home. Washington, D.C.: Federal Energy Administration, 1976. 22 pp. FEA/D–76/269. Available from Consumer Information Center, Pueblo, CO 81009.
Short do-it-yourself guide to insulating already existing homes.

Insulation Guide. Prepared and distributed by the Electric and Gas Utilities of New Jersey, 1975.

In the Bank . . . or Up the Chimney. Available from the Superintendent of Documents, U.S. Government Printing Office, Washington, DC 20402; order number 023–000–00297–3; $1.85.*
A dollars and cents guide to energy-saving home improvements. A graphic guide for the do-it-yourselfer; a real bargain at the price. Probably one of the best done and most useful of all government publications.

Low-Cost, Energy-Efficient Shelter for the Owner and Builder. Edited by Eugene Eccli. Emmaus, Pa.: Rodale Press, 1976.

*Please contact source for latest price before ordering as prices subject to change.

Brings together the information and design principles necessary to build sensibly and at an affordable price. Includes chapter on insulation.

Making the Most of Your Energy Dollars in Home Heating and Cooling.   Madeleine Jacobs and Stephen R. Petersen. U.S. Department of Commerce, National Bureau of Standards, 1975. 17 pp. Available from the Superintendent of Documents, U.S. Government Printing Office, Washington, DC 20402; order number 003–003–01446–0; $.70.*
A "how much" guide to energy conservation investments in the home.

Measures for Reducing Energy Consumption for Homeowners and Renters.   Office of Energy Systems, Federal Power Commission, Washington, DC 20426. March 25, 1975. Available from National Technical Information Service, Springfield, VA 22161; order number PB 240–472.

Program to Evaluate and Demonstrate Conservation of Fossil Fuel Energy for Single-Family Dwellings.   Naval Weapons Center, China Lake, CA 93555. June 1975. Available from National Technical Information Service, Springfield VA 22161; order number PB–245–064.

Project Retro-Tech. Home Weatherization Manual.   Richard Hill, Charles W. Kitteridge, and Norman Smith. Washington, D.C.: Federal Energy Administration, Office of Weatherization for Low-Income. Conservation Paper. no. 28C. FEA/D–75/457 R (rev. 5th printing May 1976). Available from the Superintendent of Documents, U.S. Government Printing Office, Washington, DC 20402; order number S/N 041–018–00143–3.
Intended for use in inspecting and evaluating homes to determine if weatherization measures are needed and gives directions for applying weatherization materials.

Retrofitting Existing Housing for Energy Conservation, An Economic Analysis. Stephen Peterson. U.S. Department of Commerce, National Bureau of Standards. 76 pp. Available from Superintendent of Documents, U.S. Government Printing Office, Washington, DC 20402; order number 003–003–01360–9; $1.35.*
On the technical side—for heating engineers.

Tips for Energy Savers: In and around the Home, on the Road, in the Market Place. Federal Energy Administration, 1975. Available from the Superintendent of Documents, U.S. Government Printing Office, Washington, DC 20402; order number 1975 0–566–806.
Tips for conserving energy at home, in the car, and at work.

*Please contact source for latest price before ordering as prices subject to change.

Is Your Home Suited?

Now that you've completed your basic training (chapter 2), you can see that solar heating is relatively simple. It is, in theory. Back in chapter 2 we said that *in every solar heating system* the principle is the same:

1. sunlight heats a collecting surface;
2. that surface-heat is stored somewhere, and
3. the stored heat is used when it's needed.

In taking this basic theory, and applying it to your home's unique situation, we encounter a more complex problem, however. The many variables involved with *where* and *how* you live all bear upon the question of whether or not your home is a suitable candidate for solar heating.

In this chapter we will examine the *where* by looking closely at your home and its geographic location, and will discuss the way in which each affects the building's solar potential. Next, we will question *how* you live, and discuss your life-style's impact on your home's solar potential. And, finally, we will be in a position to talk about the best solar system for your home; one that uniquely fits your needs.

Although the subject matter of this chapter is obviously directed toward the present-day homeowner who is examining his building with an eye to adding solar heat, don't despair if you are not in that category. Anyone planning to buy a home, or have one built, can benefit from the common sense and logic of the information we will be covering. In fact, we have had inquiries from realtors in New Jersey with regard to the use of solar feasibility as a sales enhancement in marketing the homes in their listings. Some realtors are already planning to use a commercially available survey* to determine which "for sale" homes have the greatest solar potential. (We wonder what they'll tell prospective buyers about homes whose score is in the lower range of the solar potential scale?)

How does *where* you live have an effect upon whether or not you would be wise to install solar heating? That question is deceptively simple. *Where* can refer to the type of house (that yellow two-story colonial down the road) or its geographic location (in the foothills of the Rockies near Denver, Colorado). The following chart briefly illustrates just how complex the question of *where* you live can be.

WHERE DO YOU LIVE
Range of answers from one extreme to the other

Type of Home			
	Large	Modest	Small
	Garage	Carport	No garage
	Masonry	Masonry/frame	Frame
	Well insulated	Some insulation	No insulation
	Full basement	Crawl space	No basement
	Multistory	Split-level	One floor
	Fully landscaped	Partially landscaped	No trees or shrubs
	Electrically heated		Gas heated
Geographic Location			
	Hot climate		Cold climate
	Wet climate		Dry climate
	Northern latitude		Southern latitude
	Always cloudy		Always sunny
	Smoky/hazy		Crystal clear
	Mountainous		Flat
	Windy		Still
	Northern Hemisphere		Southern Hemisphere

This should give you glimmerings of insight into the relationship between *where* you live and the complexity of determining your home's solar potential.

Because there are so many variables that could affect the suitability of your home's use of solar energy for heat, it is difficult to say which is the most important. For instance, if your home is shielded from the sun by a large apartment complex to the south, it would be pointless to dis-

*Found in *Your Home's Solar Potential* (see Bibliography at end of this chapter).

cuss the impact that solarizing your home would have upon your present heating system. But that's an extreme example. In most cases you will have to consider the following major factors before you decide to install solar heating:

1. your home's rate of heat-loss
2. its orientation and exposure to sunlight
3. the slope of your roof and possible areas for mounting solar collectors
4. your home's existing heating system
5. installation restrictions, both indoor and outdoor

YOUR HOME'S RATE OF HEAT-LOSS

If this were a conventional book on solar heating, we would now get into the subject of Btu's (British thermal units).* Most heat-loss calculations by engineers are in terms of Btu's, and are lengthy, complex, and (sadly) not always too precise. So, rather than waste your time and ours, and possibly lose you by boredom or frustration—or both—we will examine your home's heat-loss in a simpler way. Once you recognize the contributing factors, you can evaluate your home's heat-loss on the scale from "good" to "bad," and determine what course of action to follow.

One key element is the *effectiveness of your building's insulation.* Because of its importance, not only to your home's heat-loss, but to the total concept of sensible energy use, we devoted the entire previous chapter to this subject. For now, though, what you want to consider are the commonsense aspects of heat-loss relating to insulation. Think about questions like the following, and decide whether or not you're satisfied with the answers that come to mind. (If you're not, start making an "action item" checklist of steps you should take to correct the problems.)

- Are your heating bills greater than those of neighbors living in similar housing at a similar comfort level?
- Do you feel drafts in certain rooms in spite of your having turned up the thermostat?
- Does the snow melt from your roof more quickly than from

*A definition of Btu is given on page 38 in chapter 3 on insulation.

your neighbor's roof? (This generally indicates that more heat is being lost through your ceilings.)

• Are some rooms almost uninhabitable in the bitter-cold winter months, when the rest of the house is quite comfortable?

If the answer to each of these questions is "yes," do not continue to read this chapter now. Go back to chapter 3 on insulation, and find out how easily you can correct your problems. After they're corrected it's time to return to our discussion here.

Another key factor affecting your heat-loss is the *age, style, construction,* and *condition* of your building. Looking at these elements one at a time will help clarify their impact on your home's heat-loss.

• *Age:* Homes built before the era of cheap energy (before the early 1900s) were frequently quite energy efficient. Fuel supplies related to the sweat of one's brow and one's ability to chop and split firewood. There were few windows (a major source of heat-loss), and natural insulation and windbreak plantings were used prudently. People worked with nature, to the extent of their understanding. Then came the age of central heating, first with inexpensive coal, and later with cheap oil and gas. Homes built during this era (early 1900s through 1940s) made little use of insulation, and practically no thought was given to heat-loss. If evergreen plantings sheltered the northwestern sides of the houses from bitter winter winds, it was usually due to accident or dumb luck. If one were uncomfortable during particularly cold months, all one need do was push up the thermostat. The cost involved was negligible. Later, as our awareness of finite resources began to dawn upon us (mainly as a result of fuel prices rising), construction techniques began to reflect the need for insulation, and homes built in the 1950s and 1960s were modestly insulated. Only recently (in the mid- to late 1970s), have well-insulated homes begun to appear on the market. See how the *age* of your home can provide an initial clue to its heat-loss?

• *Style:* As simple a style concept as your roof overhang can materially affect your home's heat-loss. A properly designed overhang will shade windows from the high, hot summer sun, but will allow the low winter sun to shine in. Inner city row houses, attached to one another, have only the front and back walls (rather than all four) exposed to the environment. Low, one-story ranch-type homes have much less wall area exposed to brisk chilling winds than do multilevel houses. This list could continue, but the point is simply that you should be aware of the influence of your home's *style* on its heat-loss.

• *Construction:* A wood frame house has better insulating properties than a brick or masonry one. That is not to say that a stone or brick house cannot be well insulated. With proper techniques their resistance to heat-loss can be increased substantially, but such techniques were not always used when many brick and stone houses were originally built. Frame and clapboard buildings often had various types of insulated siding added at repainting time: This is a "plus" when considering heat-loss. Concrete slabs on the ground floor contribute greatly to heat-loss, permitting a home's warmth to escape to the cold earth surface and outside air through the edges of the slab.

• *Condition:* The effect of your home's condition on heat-loss is more easily recognized than any of the preceding factors. And it is often the most easily improved one as well. Take a walk around your home and carefully look for signs of disrepair. Cracked or broken windows . . . missing segments of siding . . . broken or cracked trim around windows and doors . . . gaps and spaces where they shouldn't be. If you read chapter 3 on insulation, you'll recall that every crack or opening in the shell of your building adds to the already high level of heat-loss almost every home suffers from due to infiltration.

The next contributing factor to your home's heat-loss is the number, type, and location of windows and doors. Next to the amount of heat lost through the ceilings and roof areas of homes, windows and doors are the worst culprits. Storm windows and doors go a long way toward solving this problem, but do not provide the complete answer. To get a better idea of how much heat is being lost this way, ask yourself the following questions:

• How many exterior doors do I have? Are they weather-stripped and tight fitting? Do I use storm doors? Is there an air-lock (vestibule) at each exterior door?

- How many windows do I have? Are they fixed or operable? Casement type? Bay windows? Double-hung? Picture windows? Are the large windows single or double glazed? How many are protected by storm windows?
- Is there a greater window and door area to the south (good!) and east, or to the north and west (not so good)? A great deal of wintertime heat-loss occurs from infiltration due to the prevailing northwest and northerly winds in most of the United States.

After reviewing these questions, note on your "action item" checklist the actions you can perform which will improve your home's resistance to heat-loss, actions such as adding storm windows or doors . . . double glazing large window areas . . . closing off a small entrance corridor to form an air-lock vestibule at an exterior door. The payoff will be great in many ways: added comfort, lower fuel bills, conserving energy for future generations, and making your home a better candidate for solar heating.

The last key factor to be considered in evaluating your home's heat-loss is landscaping—the types of trees and shrubs planted around your home. Windbreak plantings to the north and west of your house will generally reduce its heat-loss. These should be composed of evergreens, and should be planted close to the walls. Deciduous trees (the ones that lose their leaves each year) serve well if they are primarily on the southern semicircle around your home. There, they shield the house from the hot summer sun, but after losing their leaves, they permit the winter sun to warm your building (and thereby replace some of its heat-loss with solar energy).

By now, of course, you have seen that the subject of your home's heat-loss is a complex one. At the beginning of this segment, we stated that most heat-loss calculations by engineers are lengthy, complex, and (sadly) not always too precise. It is for this reason that we have chosen to present the material for a more intuitive response than an academic one. *It is unimportant to know the exact number of Btu's lost from your home each hour if you have reviewed all contributing factors, and have taken all reasonable steps to correct your heat-loss problem areas.* "Leapfrog" the need for calculations, go directly to the source of the problem, and take corrective action.

On the other hand, if you had a heat-loss survey performed by a professional engineering organization, this is what you would learn:

- A heat-loss of 60 or more Btu's/square foot/hour of heated area would indicate that your home had a severe heat-loss problem, and that you should start checking all the factors we've noted in this segment of the chapter.

• A heat-loss of about 40 Btu's/square foot/hour of heated area would show that your home had a rather average amount of heat-loss, and that you still should start checking all the factors we've noted in this segment of the chapter.

• A heat-loss of about 20 to 25 Btu's/square foot/hour from the heated portion of your home would indicate that yours is a relatively tight house, and that you might have wasted your time in reading this part of the chapter.

YOUR HOME'S ORIENTATION AND EXPOSURE TO SUNLIGHT

A coloring book* we wrote several years ago to explain solar heating to children under 60 had one sketch which explained better than words the need for sunlight exposure:

*Tilly's Catch-a-Sunbeam Coloring Book.

How else can we say it? If your home's exposure to sunlight is greatly diminished by adjacent buildings, hills, mountains, trees, or any other natural or man-made obstruction, there is very little you can do except to concentrate on energy conservation techniques. If you're reading this during summer months, don't forget that the winter sun appears much lower in the sky than does the summer sun, and as a result, distant buildings, hills, and trees, may cast their long shadows on your house during the cold winter months.

Distant buildings cast long shadows.

For most of you, though, this is not a problem, so we will look more closely at your home's orientation, and its effect upon your home's solar potential. If your home were round, there would be no concern about its orientation with respect to the sun, because its solar exposure, no matter how it was placed on your property, would be the same. This is not true of most rectangular homes. Generally speaking, they are built on an axis which coincides with the ridge of their largest roof area.

Take a moment here to better understand why we are interested in this axis. There is simply a need to know where solar collector panels will have the greatest exposure to sunlight, and therefore do the most

A round house.

effective job of heating your home. Although this reason is simple and straightforward, there are several subtle considerations you must ponder:

• As the earth rotates, the sun rises in the east, moves through the southern sky, and sets in the west. To use this information, you ob-

Magnetic compass.

viously need to know where south is with respect to your home. Three different ways to accomplish this are—

1. By using a magnetic compass. (Don't forget that magnetic south can vary from true south by as much as 8° to 10°.)

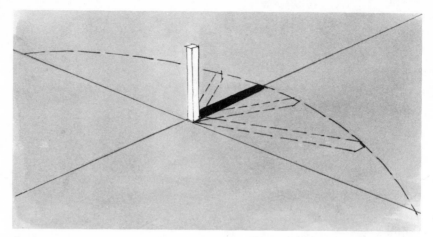

Shortest shadow is north-south line.

2. By observing a shadow cast by an upright pole or stake. When the shadow reaches its shortest length (sometime around noon), mark the spot, and the line between that spot and the base of the upright is a true north-south line.

3. By locating your original property survey that you received when you purchased your home. Somewhere on the survey will be a north arrow.

Only when you know where south is can you evaluate how your home's orientation affects its solar potential.

• How far off true east-west is the axis of your home? (90° off is the maximum deviation.)

Maximum exposure to sunshine will occur if the axis of your home is on an east-west line. In that way the sun during its full morning-to-evening cycle will shine on the walls and roof of the south side of your home. If the ridge of your roof is on a north-south axis, one side will get the morning sunshine, and the other the afternoon's. Since we are interested in having solar collectors mounted so they get maximum exposure to sunshine, an east-west axis with its all-day potential provides more wall or roof space with which to accomplish the task efficiently.

• A home axis deviation of about 15° from an east-west line will have very little effect upon the efficiency of a proposed solar heat-

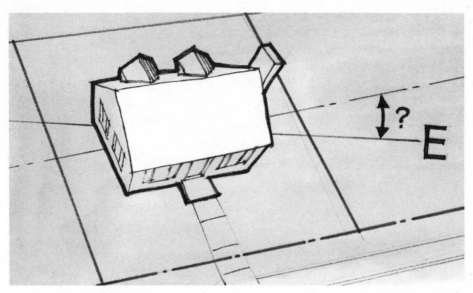

How far off the east-west line?

ing system. However, as that deviation angle increases, your hopes for effective solar heating must diminish. The home with its axis on a north-south line has, of course, the least potential for an effective solar heating system since collectors mounted on its largest roof area or wall areas would be exposed to sunshine only in the mornings or afternoons. Still, you might be able to arrange efficient collectors on some south-facing surface.

• Is your home sheltered on the north and northwest with protective evergreen plantings or by a nonheated portion of the house (such as a garage or storage room)?

• Is your window area smallest to the north and northwest, and greatest to the south?

• Are there ample clear wall and roof areas on the south exposed side of your building, or are these surfaces broken with dormers, windows, and various planes—which would cast energy-robbing shadows on solar collector panels?

Once again, the commonsense questions which can be asked are many, but they all point in one direction: maximum sunlight. After determining the side of your home with the fullest and longest exposure to sunshine, see if there is logical space for mounting solar collectors . . . space adequate for the *area of collectors* which you will need (more about that in the following segment of this chapter) and space which, when adorned with solar collectors, will not destroy the general appearance of your home? Your answers to these questions will provide more food for thought relative to your home's solar feasibility.

ROOF SLOPE AND COLLECTOR MOUNTING AREA

The third of the five major factors to be considered before deciding on solar heat for your home is that of its roof slope and area. The ac-

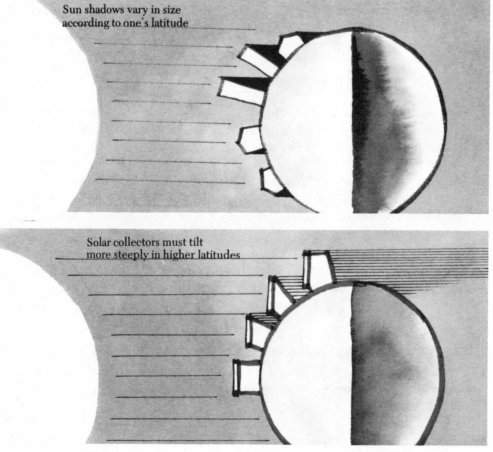

Sun shadows vary in size according to one's latitude

Solar collectors must tilt more steeply in higher latitudes

Perpendicular to the sun's rays.

companying sketch shows that we must vary the slope of sunlight collectors (relative to the earth's surface at various latitudes) if they are to catch the sun's rays in the most efficient manner, i.e.: perpendicular to the face of the collector. Fortunately, a *constant relationship* exists between the optimum (best) slope for a solar collector, and the geographic latitude of the building being evaluated. Although there is a large seasonal (summer-winter) variation in the angle at which the sun's rays strike the earth, we need consider only the average winter sun angle for heating purposes. Rather than go through the mathematics* of the formula, we'll simply state it: *For heating purposes, solar collectors will provide maximum efficiency if they are mounted at an angle of 12° + your latitude (in most temperate zones) from a horizontal surface.* In other words, if your home is located at 35° n. latitude (see any map or world globe to determine your geographic latitude), a collector angle of 47° (35+12) would provide the best wintertime solar heat efficiency. If you live at the Denver-Philadelphia latitude (40°) a slope of 52° would be best. The precise angle is not terribly important, and a variation of up to 10° in either direction will not appreciably degrade your system's efficiency.

Now that you know the best angle for mounting solar collectors, the importance of your roof slope becomes obvious. The easiest solar panel installation (fewest problems and least cost) occurs when the collector panels are mounted flat on (or near) the surface of your roof. Depending on the slope of your roof, this can provide efficiency that varies anywhere between great and horrible. So, to see whether you can enjoy advantages from solar collectors mounted at the angle of your roof, two questions must be answered:

> 1. What is its slope (and will that slope be anywhere near your northern latitude +12°)?
> 2. Is there enough clear area on the sunny side of the roof to accommodate the area of collectors needed for an efficient solar heating system? (More information about collector area follows a bit later.)

There are several easy ways to determine the slope of your roof. Some homes have several roofs at different slopes, so be sure to measure

---

*For those interested in the development of this 12° plus latitude figure, see the end of this chapter.

that roof area with the best orientation for collecting sunshine and with the greatest clear area for mounting collectors.

Folding rule remembers angle.

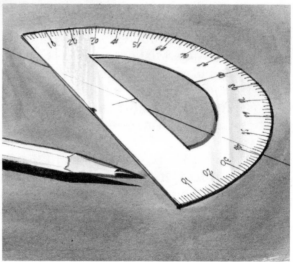

A common protractor.

• One way is to go into the attic area and place one segment of a folding carpenter's ruler flush against the underside of a roof beam, while the body of the rule is vertical. Then, using a common protractor (available from your stationery store, five-and-ten, or school supply outlet), carefully measure the angle. Repeat the procedure once or twice to be sure the angle is accurate.

• If the attic portion of your home is inaccessible, the same measurement can be taken by gently pressing the folded segment of the ruler against the underside of the roof trimboard, while keeping the long part of the rule as straight up-and-down (vertical) as possible. Repeat this process until you are confident that the end segment has the same slope as your roof, and then carefully lower the rule and read the angle from the protractor, as you did in the previous instruction.

• If neither system is convenient at your home, you can "eyeball" the roof slope. It can be done using the same carpenter's rule as shown in the accompanying sketch. Being only an approximation, this is not the most desirable technique, but, if all else fails, it will put you in the right ball park.

It's likely that your roof slope is not optimum for your latitude. (Other than those homes designed for solar heating, few in any community have just the right roof slope for solar heating purposes.) Don't let this discourage you too greatly. Again, there is tolerance for up to at least 10° in either direction from the optimum slope, without appreciably degrading your system's efficiency. If your roof slope is more than 10° from the optimum slope for your latitude, the collectors can be mounted at an angle other than that of your roof or wall, without too great an additional expense (variations of these mounting techniques were shown in chapter 2, Basic Training).

Your next concern relates to the surface (roof or wall on the southern exposure of your home) upon which a solar collector can be mounted. Several questions arise such as, "How will it look there?", "Where will the pipes and ducts run?", "Is it accessible for maintenance and repair?", but the most important question to consider now is, "How great an area of collectors will you need to heat your home?" A tough one to answer accurately.

Way back in the early days of solar heating (before 1975) a general rule of thumb was that the collector area needed to provide about 75% of your annual heating needs was half the heated area of your home.

Assuming you have about 1,800 square feet of heated area in your home (discounting nonheated areas such as garage, basement, storage areas, etc.), you would, under the old rule of thumb, need about 900 square feet of collector area. More recently, because of the many refinements in the design and manufacture of collectors, manufacturers are recommending that only one quarter to one third of your home's heated area is sufficient to provide about 75% of your heating needs. The same 1,800-square-foot home we just mentioned therefore now requires only 450 to 600 square feet of collector area. A few points of caution here:

• "75% of your annual heating needs" is a conservative goal. In mild winter months, most well-designed solar systems provide all the heat needed for a well-insulated home. During some mild winters, little or no supplementary heating is required. During extremely cold winters (such as the winter of 1976–1977) most well-designed solar systems *will* require supplementary backup heating. The reason for this is that the cost of designing and building a solar system with the capacity to accommodate the most bitter weather possible in a particular geographic area would be prohibitively high. This is because a system capable of meeting the very worst climatic conditions would be *overdesigned* for 95% of the winter days. This would be similar to having an automobile built with a diesel train engine to handle the rare need of towing a heavy truck up a steep hill.

• In measuring your heated area and estimating the necessary amount of solar collector area the "one-third of your heated area" rule of thumb is fine for rough estimating, but make your contractor do the final specifying. In chapter 7, Contracts and Contractors, we offer suggested methods of assuring a balanced design for your system, with minimum headaches for you.

In summary then, you are looking for an area of about one-third that of your heated living area; either on your south-facing walls, your roof, or both. If you can accommodate this need, your prospects for solar heat are getting brighter. In addition, if your roof slope approximates your latitude plus 12°, they are brighter still.

YOUR PRESENT HEATING SYSTEM

The next-to-last subject (whew!) to be reviewed in determining your home's solar potential is that of your present heating system. The type of heating system you presently use in your home affects the complexity (and therefore the cost) of any proposed solar heating installation. Here are some of the more common situations, presuming that you use some type of central heating system. (The cost of a solar system for those of you with local [room] heating systems will be appreciably higher because of the need for installing a new heat distribution system throughout your home—ductwork, pipes, etc.)

• If your central heating system is of the forced warm-air type, your solar installation costs are likely to be minimized. Your existing heat (or air-conditioning) distribution ductwork will probably accommodate your new solar system, and your present heating system can serve as the severe-weather backup. Either a solar hot water system or solar air system can function successfully with your heat distribution network.

• If you have a hot water heating system (baseboard radiators, conventional radiators, radiant heating in a cement slab, etc.), water temperature in the 180° to 200°F. (82° to 93°C.) range is required for effective operation. Since most solar hot water systems provide water at temperatures up to only 140° to 150°F. (60° to 66°C.) your solar installation can function as a preheater saving most of your fuel needs, but cannot supply all of the heat, even in relatively mild winter weather.

• If your heater for domestic hot water is separate from your main heating system it will accommodate either a hot water solar system (more efficiently) or a hot air solar system (less efficiently), functioning as a preheater for your hot water. Don't lose sight of the fact that your domestic hot water needs (dishes, clothes washing, bathing, etc.) account for anywhere from one-fifth to one-third of your annual heating fuel usage, tending to be higher if there are children in the family.

Still with us? We hope so. Assuming that you are still keeping track of your home's solar potential, let's take a commonsense look at your present heating system and add or subtract points based upon what you've learned so far. Keep in mind the types of solar heating systems shown in chapter 2, Basic Training.

INSTALLATION RESTRICTIONS

If the prospect of your home's solar suitability is less certain now than when you started this chapter, remember that it may well be because of your increased understanding of the many variables involved in retrofitting an existing home with a solar heat system.

We have explored the commonsense aspects of—

1. your home's rate of heat-loss
2. its orientation and exposure to sunlight
3. the slope of your roof and possible mounting space for solar collectors
4. your home's present heating system

The final element to be considered is that of *installation restrictions.* Once again we can only provide you with food for thought rather than specific guidelines, since the number of possible inside- or outside-your-home restrictions which might be encountered at various locations is mind-boggling.

The additional weight of collector panels.

Indoor Restrictions

These are some of the major concerns you must consider:

• The solar collectors: They are generally 4 by 8 feet and about 3 to 5 inches thick, and weigh as much as 200 pounds each. They are mounted outdoors, but they require connections to the inside of your home. Do you have easy access to your attic? Can your roof be penetrated easily? Will it support the additional weight of ten or fifteen collector panels?

• How about your walls: Can water lines (or air ducts) be run from the collector mounting area through your walls to your energy storage device and existing heating system (usually from the attic to the basement)? Walls may have to be broken open and rebuilt if you want to avoid exposing pipes or ducts in the living area of your home. With careful planning, wall disfigurement can often be minimized by running the piping and ductwork through nonliving areas such as attic spaces, closets, garages, and workrooms.

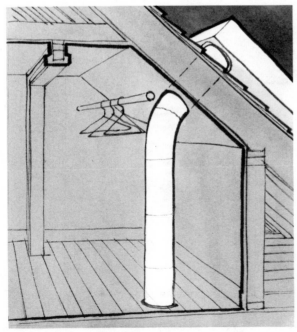

Hide the pipes and ductwork!

• An energy storage tank or bin must be located either inside, beneath, or outside (and near) your home. You may recall from reading chapter 2 that the insulated storage tank needed for an average-size hot water solar system is about the size of a Volkswagen "Bug"; about 2,000 gallons, or a volume of 250 cubic feet. An insulated rock-storage bin of similar capacity for a hot-air solar system would require about three times that volume, or about 750 cubic feet. You may also recall from chapter 2 that three times the volume *does not* mean three times as long, or as high or wide. If you look at two cubes, one being three times the volume of the other, the larger one appears perhaps 1½ times as large.

Well, this storage unit must be installed in, under, or near your basement or crawl space. Don't panic, some energy storage devices have been successfully installed in unused portions of garages and workrooms. Can this storage device be installed in your basement? Can the materials be brought in? How? Would part of the supporting basement wall have to be removed? Would there be access space for air ducts or pipes which must be routed from the collectors and heating system to this storage area? Is there room for installers and repairmen to bring in tools, and work?

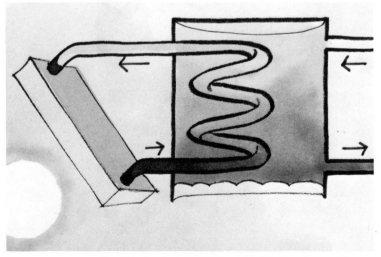

Liquid-to-liquid heat exchanger.

• A plumbing connection to your present domestic hot water system will also be necessary if you plan to use solar energy to supplement it. Is there space available, or will your domestic hot water system have to be relocated? Will new pipes to and from the system cause a problem? In solar-heated domestic water systems, it is necessary to keep the potable water clearly separated from the solar system fluid. This requires the use of a liquid-to-liquid heat exchanger, and additional plumbing and hardware, with their potential for additional problems.

• If you have been considering a liquid solar water system for your home's space-heating needs, you will also require a liquid-to-air heat exchanger similar to an automobile radiator installed in your present heating system's warm air ducts. Are they accessible for this type of modification?

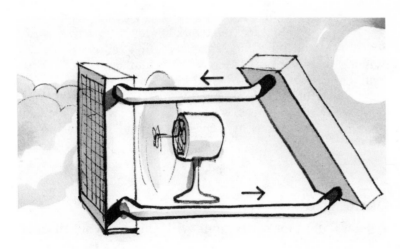

Liquid-to-air heat exchanger.

• About the easiest aspect of the installation is the electrical wiring needed for the controls, pumps, and fans. This is similar to the wiring used on conventional heating systems between the thermostat and control boxes and relays, and all wiring can readily be "fished" through walls, and installed with a minimum of household disruption.

In evaluating your indoor installation obstructions it might be wise to "keep score" as you review the items and questions previously noted.

It might be helpful to judge the difficulty of an indoor installation, on a scale of one to ten, in a manner similar to this:

Least difficult installation   **Most difficult installation**

1	2	3	4	5	6	7	8	9	10

Scores in this range reflect few obstructions to installation. Energy storage area is in an easy-access, unobstructed basement or garage, etc. Plumbing and ductwork can be routed easily. Roof is accessible for mounting collectors. Control wiring can readily be fished through walls, etc.

This range of scores would indicate average installation complexity involving some wall-opening for ductwork and plumbing. Normal basement space is available for either fluid or rock energy-storage areas. Roof area is reasonably accessible.

Scores in this range indicate rather difficult installation problems involving major tear-down and reconstruction effort on walls, foundation, etc. Difficult roof area (poor access or high). Masonry walls with no available room for inside-the-wall plumbing or duct work, etc.

OUTDOOR RESTRICTIONS

A look at potential *outdoor* restrictions will complete your installation evaluation. These restrictions relate to your property's landscaping, paving, subsoil, utility pipes, and wires. Important considerations are:

• Can vehicles gain access to your property for delivery of equipment and supplies, and for later maintenance? Some heavy hoisting equipment may be required to place collectors on the roof and to move heavy water-storage tanks into position.

• Must trees, plantings, and fences be removed in order to allow this equipment to pass?

• Are there any underground utility supply lines (gas, water, drainage, telephone, lawn sprinkler system, etc.) which might be damaged by heavy vehicles passing over, or by the excavation re-

Must trees and plantings be removed?

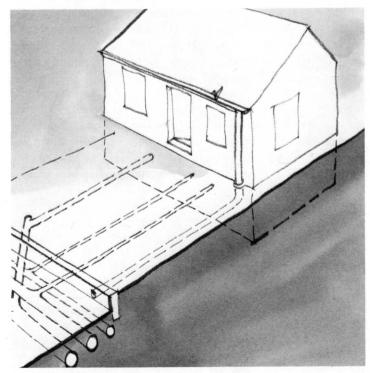

Be aware of underground utility supply lines.

quired for underground energy storage tank installation? Where are they located? How deep?

- Will overhead wires (power or telephone) restrict the movement of workers or equipment?

- Do you have plantings and trees close to the walls of your home, where they might interfere with installation work-space needs?

Once again, to help quantify your thinking, you might imagine the difficulty on a scale of one to ten:

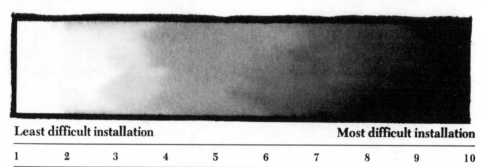

Least difficult installation   **Most difficult installation**

1	2	3	4	5	6	7	8	9	10

Scores in this range indicate an easy installation. Home is relatively free of landscaping, trees, shrubs, and fencing. Access for construction equipment is unlimited, and underground (or overhead) utilities pose no problem. Easy access to basement for storage tank location and relatively low easy-to-work-on roof.

This range of scores indicates average installation complexity. This might involve the removal of a section of fencing or some shrubbery. The difficulty of a roof-mounted collector installation would be about average. Underground and overhead utilities may have to be temporarily rerouted.

This range of scores relates to difficult installation situations. Heavy shrubbery and plantings must be removed. Possibly the removal of several trees. Concrete paving around home may have to be removed and replaced in an area for burying the heat storage tank. No practical way for construction equipment to gain access to the property.

Now that you've had the opportunity to review the commonsense considerations relating to your home's solar suitability, what are you going to do about it? Facing up to reality causes strange reactions in some people; the following example may indicate how extreme these reactions can be. (This is an actual situation which occurred early in 1977.)

We were contacted by an attorney in April 1977 shortly after President Carter's first energy message. We were retained as consultants and were asked to determine the solar feasibility of the home the lawyer was about to have built by a local contractor. The lawyer was a man who felt strongly about conservation and had decided that his new home should be solar heated. Since the plans were already drawn, we requested a set of prints in order to check the design's solar potential.

A glance at the site plan showed us that the home was to be built on a cul-de-sac as indicated in the following sketch. The most obvious

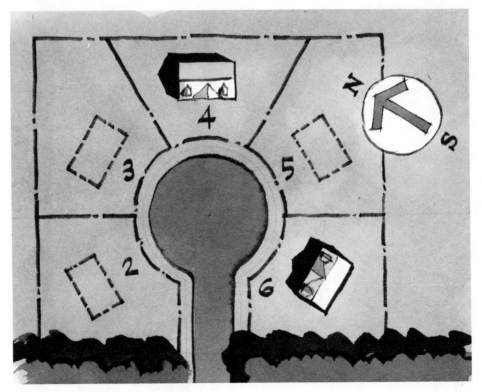

The home was to be built on a cul-de-sac.

fact to greet us at first glance was that the house would face the wrong way, and that roof-mounted collectors would get only morning or afternoon sun exposure. In addition, the front elevation—the western exposure—indicated an "English Tudor" facade of brick, with small windows—while the eastern exposure—the rear of the house—was quite contemporary, having substantial glass area.

The fake English Tudor facade.

We asked the lawyer if all other parcels on the cul-de-sac were sold—if he might possibly switch his house to lot #6, with its many attendant benefits. The new location would present a wide expanse of unbroken roof area (in the rear of the house) to a southern exposure, with maximum potential for collecting solar energy. In addition, the large window area on that side would act as a natural solar collector, further enhancing the conservation aspects of the house. Facing northeast on his current site, those same rear windows would cause a continuing heat-loss. Simply by moving it from lot #4 to lot #6, the house would go from being a "poor" to a "good" solar candidate. And, as luck would have it, lot #6 was still unsold, and the contractor was willing to switch sites at no extra charge.

The lawyer said he'd check with his wife and let us know.

After waiting a few days for his decision and for the go-ahead on certain drawing modifications needed to increase the home's solar potential, we received a call from the attorney. He said he'd decided to stay on site #4, and not to solarize the house! Surprised and disappointed that another good solar possibility was about to be lost, we asked what had caused him to change his mind. His answer was even more distressing. He and his wife felt that since they were spending $10,000 extra for the fake English Tudor facade, the house should face the street entrance to the cul-de-sac. If they switched to site #6, he explained, their expensive facade might not be as visible to those driving by. So much for his commitment to energy conservation.

As we said earlier, facing real choices causes strange reactions in some people. And because everyone reacts individually and personally to various problems, you must now examine your own reaction to the way *you* live. This chapter has explored the commonsense aspects of *where* you live; now you must consider *your life-style's impact* on your home's solar potential. Then you'll be able to make a reasonable judgment as to the best solar system for your particular needs.

While this chapter is fresh in your mind, select a ranking on the following bar chart that best describes your *home's* solar potential. Keep it in mind because we'll be using it when you get to the end of the next chapter, to help determine the best solar system for your needs.

HOME

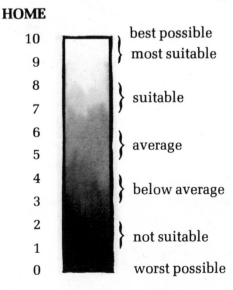

10	best possible
9	} most suitable
8	
7	} suitable
6	
5	} average
4	
3	} below average
2	
1	} not suitable
0	worst possible

APPENDIX /CHAPTER 4

Choosing the Best Angle

The three charts here show the height of the sun measured at 40° north latitude at each hour of a day on—

1. the longest day, June 21

	6	7	8	9	10	11	12	1	2	3	4	5	6
					← AM		NOON		PM →				
DEGREES	15°	26°	37.5°	49°	60.5°	69.5°	73°	69.5°	60.5°	49°	37.5°	26°	15°

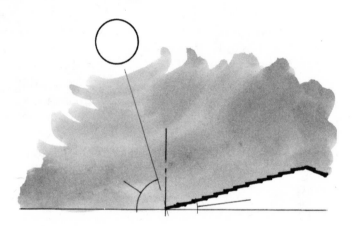

2. the shortest day, December 21

	6	7	8	9	10	11	12	1	2	3	4	5	6
					← AM		NOON		PM →				
DEGREES	—	—	5.5°	14.5°	21°	25°	26.5°	25°	21°	14.5°	5.5°	—	—

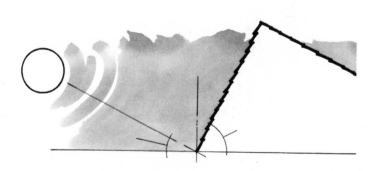

3. a mid-season day, March 21 (or September 21)

	6	7	8	9	10	11	12	1	2	3	4	5	6
					← AM		NOON	PM →					
DEGREES	—	11.5°	22.5°	33°	41.5°	48°	50°	48°	41.5°	33°	22.5°	11.5°	—

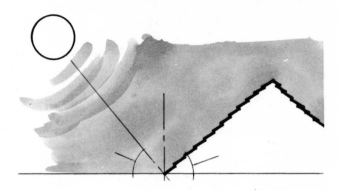

In each case the maximum height of the sun from horizontal is shown in degrees. Perpendicular to that highest sun angle is shown the "best" angle for a solar collector, on that particular day. This "best" angle is really the angle at which the most direct rays of the sun will be vertical to the collector when the sun is at its peak.

As you can see, the angle varies considerably. If we wanted the greatest amount of energy *only* on the shortest day of the year, we would want a collector slope of 63.5°. If we wanted an average slope for year round (summer, winter, fall, and spring), we would use a slope of 40°. But, since we're really interested in maximizing the energy collected during the heating season (late fall, winter, early spring) we average the two readings in the following manner:

Season or Date	Reading
Shortest Day—December 21	63.5°
Mid-Spring /Fall—March /September 21	40.0°
	103.5 ÷ 2 = 51.7 or 52°

This same calculation, when performed with sun heights measured at various latitudes, consistently provides an optimum slope of the latitude plus about 12° in the Northern Hemisphere.

BIBLIOGRAPHY

Buying Solar. Joe Dawson. Federal Energy Administration. Available from the Superintendent of Documents, U.S. Government Printing Office, Washington, DC 20402. Order number 041–018–00120–4.
Rather formal, but filled with many useful tables, charts, and indices.

Energy in Building Design. Davis and Shuberts. October 1974. Virginia Polytechnic Institute, College of Architecture.

Energy Conservation and Solar Retrofitting for Existing Buildings in Oregon. 1975. The Center for Environmental Research, School of Architecture and Allied Arts, Eugene, OR 97403.

Energy Primer. Whole Earth Truck Store and contributing authors. 1974. Portola Institute, Menlo Park, CA 94025.
Looks like the Whole Earth Catalog—jam-packed with all kinds of information on solar, water, wind, and biofuel energy conversion.

Here Comes the Sun. May 1975. Joint Venture and Friends, Boulder, CO 80302.

"How to Make Sunpower Work for You." Edwin Kiester, Jr. *Skeptic,* March/April 1977, pp. 49–53.

Solaria. Homan, Thomason, Wells. Published by Edmund Scientific Company, 101 East Gloucester Pike, Barrington, NJ 08007. April 1975.

Solar Dwelling Design Concepts. AIA Research Corporation for HUD, office of Policy Development and Research. May 1976. HUD document HUD-PDR-156. Available from the Superintendent of Documents, Government Printing Office, order number S/N 023-000-00-334-1; $2.30.*
Very graphic–an excellent document from which to learn and understand the basics involved in solar energy.

The Solar Home Book. Bruce Anderson. Harrisville, NH 034501: Cheshire Books, 1976.
One of the best reference texts available. Somewhat more technical than the average homeowner might need, but loaded with worthwhile information.

Tilly's Catch-a-Sunbeam Coloring Book. Tilly Spetgang. Solar Service Corporation, 306 Cranford Road, Cherry Hill, NJ 08003. 1975.

*Please contact source for latest price before ordering as prices subject to change.

Your Home's Solar Potential. Malcolm Wells and Irwin Spetgang. Edmund Scientific
  Company, 101 East Gloucester Pike, Barrington, NJ 08007. 1975.
  Just noting the authors of this book tells you something about it.

5

Is Yours a Solar Family?

There's a direct relationship between conservation and the successful application of solar heating.

This theory isn't startlingly original; it's just common sense. If yours is a concerned family, if you want your children to enjoy the same energy benefits you've had, then you *must* be a conservationist! The question is . . . how far do you go?

We've compiled a list of questions exploring many facets of conservation. Keep a tally of your answers, and see if you emerge with a majority of "yes" replies. If so, solar installation may completely satisfy you since you'll not be making impossible demands on it.

But if you come out with a majority of "no" replies, perhaps it's time for you to think about adjusting your life-style. Not only will the adjustment help you attain success in the application of solar heat, it will help this tender blue green planet's chances for a blue green future.

Some of the questions to be asked are quite pointed, and may cause you to wrinkle your brow in self-examination. Don't worry if many of your answers are negative; even we, the wonderful authors, cannot answer all the questions with a positive response. The questions are in five categories:

Heat
Hot water
Electricity use
Travel
Wastes

HEAT

Controlling the temperature of the space you live in represents about 60% of the total energy used in your home.

	Yes	No
1. Do you keep your home's thermostat set no higher than 70°F. (21°C.) during the heating season?	____	____
2. Have you attempted to reduce the setting by 2°F. (1.11°C.) per year during the heating season?	____	____

Most people soon learn to tolerate, and then prefer, a daytime temperature of 64° to 68°F. (18° to 20°C.). Wear a light sweater, or other appropriate clothing for comfort during particularly cold spells.

3. Do you set back your thermostat an additional 5° to 10°F. (2.78° to 5.55°C.) at night, and in the daytime when the home is unoccupied?	____	____
4. Have you improved your home's heat-loss/heat-gain characteristics by improving its insulation as outlined in chapter 3?		
Priority 1—above-ceiling insulation?	____	____
Priority 2 — window and door treatment?	____	____
Priority 3 — weather stripping and caulking?	____	____

Insulation and weather stripping provide benefits in both the heating and cooling seasons.

5. Are evergreens, or other windbreak plantings growing close to your home on the northern and western exposures, or providing you protection from whatever direction your prevailing cold winter winds come?	____	____
6. When a room of your house is not used, do you close off its heating and cooling?	____	____
7. Have you added air-locks (vestibules) at all the exterior doors of your home?	____	____
8. Has your heating and cooling system been maintained properly?	____	____

Some homeowners do this themselves, but if

you're unsure of the tune-up methods, frequency of replacing filters, etc., have a professional do the job for you. Both your heating and cooling systems should be checked over at least once a year; filters should be replaced much more frequently. It's also important to have your air ducts professionally cleaned every five years or so. Clean ductwork provides several advantages: less restriction to air-flow since cleaner ducts result in cleaner filters, less dust circulating in your home, fewer objectionable odors from the heat ducts . . . just to name a few.

9. Have you checked with your local heating contractor about new or improved burner elements that can be added to your furnace? ____ · ____

10. Have you installed any heat-saving devices to recover wasted heat that otherwise goes up your chimney? ____ ____

11. Do you control the draft caused by your fireplace chimney by closing the damper when the fireplace is not in use? ____ ____

12. Have you considered using a ceiling fan during the *heating* season? ____ ____
    Such fans use relatively little energy and can reduce fuel bills by moving warm air from the upper reaches of your rooms back down to where it's needed.

13. Does your home have a humidifier hooked into the heating system? ____ ____
    A relative humidity of 40 to 50% in your home during the heating season offers many benefits: Furniture doesn't dry out and crack, or loosen at glued joints. Dried-out nasal passages are the cause of frequent infections, and many doctors advise that more humidity is more healthful in the winter. Also, lower wintertime temperatures feel more comfortable when humidity is at the proper level.

14. Is your air conditioner thermostat set no lower

than 78°F. (26°C.)?   ____ ____

15. Have you tried raising the thermostat 1°F. per year during each air-conditioning season?   ____ ____

This note is for our readers in the mid- to upper United States. Because your summers aren't as intensely hot as those in the southern regions, you might try keeping your air conditioner turned off, except for the few true "dog days" each summer. For many of you who use the air conditioning because of the allergy sufferers in your family, have you looked into the possibility of desensitizing shots which your doctor can provide? When they work, you have the added advantages of open windows, fresh air, birds singing, summer fragrances, and the delightful quiet that comes from the air conditioner being off.

16. Have you installed sunshades or screens on south-facing windows, to shelter you from hot summer sun exposure?   ____ ____

17. Do you minimize the use of heat-generating electric lights and appliances during the summer season?   ____ ____

18. Have you provided a summer sun screen to the south of your home?   ____ ____

The sun screen may be a trellis arrangement with slats set at an angle to shut out the high summer sun and admit the sun's winter rays, or simply shade trees or other deciduous plantings.

19. Have you checked the outside unit of your central air conditioning, or the outside exposure of your window units, to be sure that weeds, shrubs, or trees are not restricting the flow of air?   ____ ____

HOT WATER

Your family's use of hot water represents about 20% of the total energy used in your home.

1. Have you reduced the temperature setting on your hot water heater to 140°F. (60°C.) or less? _____ _____
   Many home hot water heaters are set as high as 180°F. (82°C.), or even higher. This is excessive and wasteful. (Water at only 110°F. (43°C.) feels hot to the touch.)

2. Have you investigated or installed water-saving faucet and shower heads for your kitchen and bathrooms? _____ _____
   These devices offer the added advantage of conserving water, as well as the energy needed to heat it.

3. Have all leaking faucets been repaired in your home, particularly on hot water outlets? _____ _____

4. Have you insulated your hot water pipes? _____ _____

5. Do you plan your laundry work so that only full-load use is made of your automatic washing machine? _____ _____

6. Do you ever think twice, and sometimes reconsider, before placing an article of clothing into the dirty laundry hamper? _____ _____

7. Do you follow the manufacturer's recommended control settings on your washing machine? _____ _____
   Review the owner's manual which came with it. Avoid using hot water when warm will do the job. Use cold water for laundry whenever you can.

8. Do you clean the lint and debris traps on your washing machine and dishwasher as a routine matter? _____ _____
   This permits the equipment to function more efficiently, thereby using less energy.

9. Are you one who does not permit the hot water to run needlessly when rinsing dishes to load your automatic dishwasher? _____ _____

10. Do you run your dishwasher only for full loads? _____ _____
    Keep extra glasses, saucers, cups or whatever on hand so that you won't be inclined to run a partially empty dishwasher when you run short

of a particular item.

11. Do you take brief showers rather than lengthy ones
    or baths?   ____ ____

12. Have you tried turning off the hot water when
    lathering your body in the shower?   ____ ____

13. Do you bathe or shower infrequently?   ____ ____
    For those whose work is dirty, and for hard-play-
    ing children and athletes, it may be necessary to
    bathe daily. For most adults though, the daily
    bath or shower is an energy-consuming bad
    habit.

14. If there are several small children in the house, do
    you bathe them together?   ____ ____
    This is not only fun for the children, but it saves
    hot water, and can save wear and tear on Mom
    and Pop.

15. Is the capacity of your hot water heater consistent
    with your needs?   ____ ____
    Heating and storing unused gallons of hot water
    is wasteful. If your children are grown and the
    hot water demands of your household are
    reduced, consider installing a smaller, well-
    insulated hot water heating unit, and make sure
    it's located near the place where the water is
    used. In a very large home it pays to have small
    water heaters near the places where hot water is
    needed.

ELECTRICITY USE

We've already questioned you about your dishwasher and automatic washing machine, but how about all the other appliances?

1. Do you hang clothes outside to dry when the
   weather permits?   ____ ____

2. Do you set the timer on your clothes dryer for the
   shortest cycle to dry your laundry effectively?   ____ ____

3. Are you upset when you find lights burning need-
   lessly in unused areas or rooms?   ____ ____

4. Do you turn them out; even if it's not your home where the waste is occurring?   ____ ____

5. Are your walls and ceilings painted with light colors to provide the maximum reflection of natural and electrical lighting?   ____ ____

6. Have you minimized the use of general overhead lighting by using task lighting where possible?   ____ ____

7. Have you cleaned the reflecting surfaces and globes of your lighting fixtures lately?   ____ ____

    This can reduce the need for higher-wattage, energy-wasting bulbs, where lower-wattage bulbs will do the job.

8. Have you installed dimmer controls wherever possible?   ____ ____

    Beyond energy savings, they provide the added benefit of increasing the life of all bulbs controlled by the dimmers, and can also provide dramatic lighting.

9. Have you minimized the decorative lighting normally used at Christmas?   ____ ____

10. Has the gas company turned off any outdoor gas lamps on your property?   ____ ____

11. Do you turn off heating elements promptly after cooking?   ____ ____

12. Do you use the lowest heat setting needed to cook whatever you are cooking?   ____ ____

13. Do you match your pan size to the size of your stove surface unit?   ____ ____

    Doing this, and also putting lids on pots, results in a more efficient use of energy.

14. Do you take full advantage of your oven heat by baking several dishes at one time?   ____ ____

    Preheating your oven should be done only when absolutely necessary.

15. Do you use a pressure cooker?   ____ ____

    Pressure cookers, and the use of steaming trays, can reduce energy consumption by as much as one-third, and help retain flavor, texture, and vitamin content in most foods.

16. Do you schedule meals so that everyone in the household can be fed at one sitting? ____ ____

17. When using your oven, do you keep it closed until all cooking is done? ____ ____

    Each peek into the oven drops the temperature from 15° to 30°F. (8.33° to 16.61°C.) wasting energy.

18. Do you cook extra portions, for later meals, when using your stove? ____ ____

19. Do you own a small burner-top oven, or Dutch oven, for those small baking jobs? ____ ____

20. Are you aware of the fact that heat should be reduced after something's brought to a boil? ____ ____

21. Have you considered turning off pilot lights in your gas stove and oven? ____ ____

    As much as one-third of your cooking gas could be saved. A welder's torch lighter (flint sparker, like a cigarette lighter) is very handy for lighting your stove, and less dangerous than matches.

22. Do you eat as little meat as possible? ____ ____

    This is generally more healthful for you, and it means that far less fuel, and land, are needed for the purpose of feeding you.

23. When using previously frozen foods, do you permit them to thaw before cooking? ____ ____

    This should be done in the refrigerator where they can contribute to the cooling effect while they are thawing, and save energy.

24. Do you defrost your freezer whenever the frost gets to be about ¼-inch thick? ____ ____

25. Do you think ahead for what you will take from your refrigerator or freezer? ____ ____

    It's wasteful to stand in front of an open refrigerator or freezer, surveying the scene, and thinking about what you might want to take out.

26. Do you periodically clean the accumulated lint and dirt from the refrigerator's condenser coils (usually at the base of the appliance)? ____ ____

    Not doing this is an invitation for poor performance from your refrigerator and freezer, as well as a waste of energy.

27. When leaving your home for an extended period of time, do you clean out and unplug your refrigerator? ____ ____
28. Do you feel that an electric can opener is unnecessary? ____ ____
29. Do you replace vacuum cleaner bags at regular intervals, so that the appliance operates more efficiently? ____ ____
30. Do you own a push-type, nonpower lawn mower? ____ ____
    If you plant a wildflower garden, as thousands have done, you can reduce your lawn area so much that hand-mowing can be done in a short time.
31. Do you have, and use, nonpower equivalents for each of your power tools? ____ ____
32. Are you disturbed by the needless waste when you find the television set playing to an empty room? ____ ____
    For an awareness of the tremendous waste of energy generated by the needless use of a TV set, compare the average wattage of various home appliances. This can be found at the end of this chapter.

TRAVEL

Although not directly tied in to your home energy consumption habits, these travel habits are part of your conservation consciousness, and thus, play a part in this chapter.

1. Do you use public transportation to get to and from work? ____ ____
2. Do you belong to a car pool? ____ ____
3. Are your family errands planned so that several can be accomplished on a single trip? ____ ____
4. Do you keep your motor vehicles well tuned and maintained? ____ ____
5. If you're fortunate enough to have a shopping area

within a mile or so of your home, do you some-
times walk, rather than drive, to the stores?   ____ ____
> Wheeled (folding) shopping carts are handy for
> this purpose, and permit you to carry a
> substantial amount of items without strain.

6. Do you have bicycles equipped with baskets?   ____ ____
7. Do you use them for light, local shopping and errands?   ____ ____
8. Should children be restricted from using power
toys such as motor bikes and snowmobiles?   ____ ____
> Aside from the energy they waste and the bad
> habits they form, all the exercise advantages of
> conventional bicycles and skis are lost.

9. Are you annoyed when school buses are traveling
half-empty while the high school students are
choking the school parking lots with their autos?   ____ ____
10. Have you voiced objections to your school board
about needless student driving to and from
school?   ____ ____
11. Are you aware that you conserve fuel by driving at
a steady speed (when possible)?   ____ ____
> Plan trips to avoid heavy traffic with its wasteful
> starting and stopping. Also, avoid driving too
> close to the car in front of you, so there's no need
> for quick starts and stops. A "jackrabbit" start is
> one of the worst offenders in using fuel un-
> necessarily.

12. Are you upset at the sight of someone in a parked
car with the engine running needlessly?   ____ ____
> Unless the driver is just pausing momentarily
> (this is usually not the case), he is doing several
> inconsiderate things:
> - Wasting fuel (that may someday be price-
>   less).
> - Polluting the air you must breathe.
> - Often he is parked illegally and causing
>   parking congestion.

13. Have you ever walked up to such a parked car and
asked the driver to turn off his engine?   ____ ____
14. Do you leave your car's windows open on hot

days, rather than run your auto air conditioner? ____ ____
> Depending on the condition of your auto air conditioner, it may cost you as much as 15% of your fuel expenditure. Another way of saying this is that you may lose as much as 50 miles travel on a tank of gas.

15. Have you considered the installation of a simple vacuum gauge on your gas-guzzler, to indicate the engine's efficiency. ____ ____
> Vacuum gauges are simple instruments that tell you how good or bad your driving economy is. They respond to your foot pressure on the accelerator under varying driving conditions, and can be purchased and installed for under $25.

16. Are you aware that air travel, on a passenger mile/gallon basis, is two to three times as wasteful as driving? ____ ____

WASTES

How do you manage them?

1. Do you take old newspapers and magazines to a recycling facility? ____ ____
2. Do you reuse glass containers when possible, and recycle the rest by taking them to a recycling station? ____ ____
3. Do you take used metal containers to a recycling facility? ____ ____
> These questions assume that your neighborhood has a recycling center. If it doesn't, perhaps you could be active in attempting to establish one.
> Community recycling centers often earn a handsome profit in addition to the environmental profit they automatically provide. If your community has not yet taken advantage of this project, it's missing a valuable opportunity.
4. Do you consistently put out fewer garbage cans on collection days than do your neighbors? ____ ____
> Even taking into account the differences in

family size, the amount (and type) of garbage disposed of by any family is a tangible reflection of its conservation habits.

5. Do you segregate potato skins, orange rinds, and other organic waste from your kitchen area for use in your garden compost pile?

    This question assumes you have a vegetable garden, not only saving much energy, but providing you with the satisfaction and pleasure of partaking of the fresh fruits of the earth. If not though, make it a point to buy *unprocessed foods!* You can still taste their natural goodness, and you reduce the energy consumed by the food processors.

6. Do you recycle your lawn clippings by using them as mulch for your garden beds?

    A 3-inch-thick covering of grass clippings will greatly reduce the growth of weeds while retaining moisture in the bed. The dried grass clippings also make an attractive garden dressing, decaying to form a rich compost layer which can later be turned into the earth.

7. Do you have a reusable shopping bag for use during your shopping trips?

    This can prevent the needless waste of having each item wrapped or bagged at the point of purchase.

8. Do you avoid the use of plastic (nonbiodegradable) materials, when biodegradable alternates are available?

    Bear in mind the base material of most plastic items is oil, and these items take much, much longer to decay in the landfills where they're dumped. The apparent advantages they offer make it hard not to use "disposable" plastic garbage and lawn bags, sandwich bags, plastic wrapping paper (as opposed to old-fashioned wax paper), plastic pop bottles, detergent containers, etc. Look for the alternates; plastics are bad news from a conservation viewpoint.

For each question we've asked in this chapter, you will probably think of at least one more that was left unasked, and that's good! The more conscious we all become of the "fat" aspects of our lives, the more comfortable we'll become when using less.

THE BEST SYSTEM

Chapter 2 briefed you on solar basics. Chapter 3 covered insulation. Chapter 4 described those physical qualities of your *home* which most directly affect the success or failure of a solar installation. And finally, in this chapter, you have examined your *family's* attitudes toward energy and natural resources.

Now, let's put it all together, and see which solar options appear most sensible for your unique home and family situation. This can be done in two simple steps.

Step 1

On the following two bar charts, select the ranking that best describes your home and family. The *home* bar chart is the same one you used at the end of chapter 4 to determine your home's ranking.

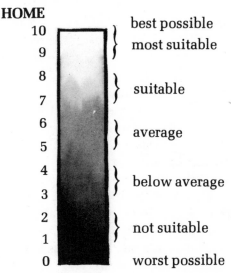

HOME

10	} best possible
9	} most suitable
8	} suitable
7	
6	} average
5	
4	} below average
3	
2	} not suitable
1	
0	worst possible

Mentally review the elements of your home which you evaluated in the previous chapter. Decide which of these rankings is most accurate for your building.

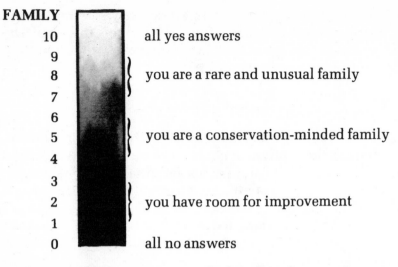

FAMILY
10   all yes answers
9
8   }   you are a rare and unusual family
7
6
5   }   you are a conservation-minded family
4
3
2   }   you have room for improvement
1
0   all no answers

With regard to the questions asked in this chapter, if you responded to less than one-third of them with a yes, you have room for improvement. If you responded to more than two-thirds in a positive way, you are rare and unusual, and have the best attitude for success in solar heating.

Step 2

Once you have your rankings on the bar graphs above, find that proper combination of them among the fifteen on the side of the following table, and you'll be all set to go. Our recommendations are indicated as shown on the table.

You'll notice that we've suggested solar heating for domestic hot water purposes more times than any other recommendation (particularly for retrofit installations). To help you pursue this option we have included a list of manufacturers of solar hot water equipment (nationwide—listed by state) in the Appendix for this chapter.

Still with us? Good. You'll be glad you didn't give up.

Now that you have an idea of how well you're likely to do in solar heating, it's time to look at how others have done, and to learn from their mistakes. The next chapter, a summary of our national poll of solar homeowners, often appears negative, but don't be discouraged by it. The poll was written around the question, "What would you (the solar homeowner) do differently if you had it to do over again today?," so naturally we got a lot of negative responses. But why don't you see for yourself. . . .

SOLAR OPTION CHART

RECOMMENDATION KEY:

X — SHOULD WORK
☆ — THE BEST FOR YOU

YOUR RANKING

YOUR RANKING	A1 Retrofit Installation — Passive Heating & Domestic Water	A2 Passive Heating Only	A3 Active Heating—Air System & Domestic Water Pre-Heat	A4 Active Heating Air System Only	A5 Active Heating—Fluid System & Domestic Water Heating	A6 Domestic Water Heating Only (Active System)	B1 New Construction — Passive Heating & Domestic Water	B2 Passive Heating Only	B3 Active Heating—Air System & Domestic Water Pre-Heat	B4 Active Heating—Air System Only	B5 Active Heating—Fluid System & Domestic Water Heating	B6 Domestic Water Heating Only (Active System)	
Home: Most Suitable / Family: Rare & Unusual	X	X	☆	X	X	X	☆	X	X	X	X	X	1
Home: Most Suitable / Family: Conservation Minded	X	X	☆	X	X	X	☆	X	X	X	X	X	2
Home: Most Suitable / Family: Room for Improvement			X	X	X	☆	X	X	X	X	X	☆	3
Home: Suitable / Family: Rare & Unusual	X	X	☆	X	X	X	☆	X	X	X	X	X	4
Home: Suitable / Family: Conservation Minded			X	X	☆	X	☆	X	X	X	X	X	5
Home: Suitable / Family: Room for Improvement			X	X	X	☆		X	X	X	X	☆	6
Home: Average / Family: Rare & Unusual			X	X	☆	X			X	☆	X	X	7
Home: Average / Family: Conservation Minded				X	X	☆			X	X	☆	X	8
Home: Average / Family: Room for Improvement				X		☆				X	X	☆	9
Home: Below Average / Family: Rare & Unusual					X	☆					X	☆	10
Home: Below Average / Family: Conservation Minded						X						X	11
Home: Below Average / Family: Room for Improvement													12
Home: Not Suitable / Family: Rare & Unusual													13
Home: Not Suitable / Family: Conservation Minded													14
Home: Not Suitable / Family: Room for Improvement													15

APPENDIX/CHAPTER 5
Manufacturers of Solar Hot Water Equipment*†

The following lists are provided for information only. The National Solar Heating and Cooling Information Center, The Franklin Institute Research Laboratories, The U. S. Energy Research and Development Administration, and The U. S. Department of Housing and Urban Development do not endorse, recommend, or attest to the quality or capability of any products or services or companies and individuals.

These lists are based on information available to the National Center at the time of publication. For an up-to-date list, write to the Center.

ALABAMA

Sun Century Systems
P.O. Box 2036
Florence, AL 35630
205/754-0795

ARIZONA

Arizona Solar Enterprises
6719 E. Holly St.
Scottsdale, AZ 85257
602/945-7477

Hansberger Refrigeration &
Electric Co.
2450 8th St.
Yuma, AZ 85364
602/783-3331

Helio Associates
P.O. Box 17960
Tucson, AZ 85731
602/792-2800

Mel Kiser and Associates
6701 E. Kenyon Dr.
Tucson, AZ 85710
602/747-1552

Sunpower Systems Corp.
2123 South Priest Rd./Suite 216
Tempe, AZ 85282
602/968-7425

CALIFORNIA

Alten Associates Inc.
2594 Leghorn St.
Mountain View, CA 94043
415/969-6474

American Sun Industries
3477 Old Conejo Rd.
P.O. Box 263
Newbury Pk., CA 91320
805/498-9700

Applied Sol Tech Inc.
P.O. Box 9111 Cabrillo Station
Long Beach, CA 90810
213/426-0127

Baker Bros. Solar Collectors
207 Cortez Ave.
Davis, CA 95616
916/756-4558

Conserdyne Corp.
4437 San Fernando Rd.

*Excerpted from "Solar Hot Water and Your Home," by the National Solar Heating and Cooling Information Center, P.O. Box 1607 Rockville, MD 20850.

†For a listing of manufacturers of solar collectors, see General Appendix at the back of the book.

Glendale, CA 91204
213/246-8409

Elcam, Inc.
5330 Debbie La.
Santa Barbara, CA 93111
805/964-8676

Energy Systems Inc.
634 Crest Dr.
El Cajon, CA 92021
714/447-1000

Fred Rice Productions
48780 Eisenhower Dr.
P.O. Box 643
La Quinta, CA 92253
714/564-4823

Grundfos
2555 Clovis Ave.
Clovis, CA 93612
209/299-9741

Helio-Dynamics Inc.
518 S. Van Ness Ave.
Los Angeles, CA 90020
213/384-9853

Heliotrope General
3733 Kenora Dr.
Spring Valley, CA 92077
714/460-3930

Kessel Insolar System
2135 Mono Way
Sonora, CA 95370
209/532-2996

Natural Energy Systems
1632 Pioneer Way
El Cajon, CA 92020
714/440-6411

Piper Hydro Corp.
2895 East La Palma
Anaheim, CA 92806
714/630-4040

Powell Brothers Inc.
5903 Firestone Blvd.
South Gate, CA 90280
213/869-3307

Ra-Los Inc.
559 Union Ave.
Campbell, CA 95008
408/371-1734

Raypak Inc.
31111 Agoura Rd.
Westlake Village, CA 91359
213/889-1500

Rho Sigma
15150 Raymer St.
Van Nuys, CA 91405
213/342-4376

Skytherm Processes Engrg.
2424 Wilshire Blvd.
Los Angeles, CA 90057
213/389-2300

Sol-Aire
46 Las Cascadas
Orinda, CA 94563
415/254-2672

Solar-Aire
82 S Third St.
San Jose, CA 95113
408/295-2528

Solar Applications, Inc.
7926 Convoy Ct.
San Diego, CA 92111
714/292-1857

Solarcoa
2115 E. Spring St.
Long Beach, CA 90808
213/426-7655

Solar Energy Digest
P.O. Box 17776
San Diego, CA 92117
714/277-2980

Solar Energy Systems Inc.
336 East Carson St.
Carson, CA 90049
213/549-4012

Solar II Enterprises
21416 Bear Creek Rd.
Los Gatos, CA 95030
408/354-3353

Solargenics Inc.
9713 Lurline Ave.
Chatsworth, CA 91311
213/998-0806

Solar Master Solar Panels
722D W. Betteravia Rd.
Santa Maria, CA 93454
808/922-0205

Solar Utilities Co.
124½ E. Cliff
Solana Beach, CA 92075
714/481-0255

Solarway
P.O. Box 217
Redwood Valley, CA 95470
707/485-7616

Solergy Inc.
70 Zoe St.
San Francisco, CA 94107
415/495-4303

Sunburst Solar Energy Inc.
P.O. Box 2799
Menlo Park, CA 94025
415/327-8022

The Sundu Co.
3319 Keys La.
Anaheim, CA 92804
714/828-2873

Sunshine Utility Co.
1444 Pioneer Way/Suites 9 & 10
El Cajon, CA 92020
714/440-3151

Sunsource Inc.
1291 S. Brass Lantern Dr.
La Habra, CA 90631
213/943-8883

Sunwater Energy Products
1488 Pioneer Way # 17
El Cajon, CA 92020
714/442-1532

Unit Span Architectural Systems
9419 Mason Ave.
Chatsworth, CA 91311
213/998-1131

Western Energy Inc.
454 Forest Ave.
Palo Alto, CA 94302
415/327-3371

Ying Manufacturing Corp.
1957 West 144th St.
Gardena, CA 90249
213/327-8399

COLORADO

Designworks
P.O. Box 700
Telluride, CO 81435
303/728-3303

Energy Dynamics Corp.
327 West Vermijo Rd.
Colorado Springs, CO 80903
303/475-0332

Enertech Corp.
R.D. 4/P.O. Box 409
Golden, CO 80401
303/642-3891

Entropy Limited
P.O. Box 2206
Boulder, CO 80306
303/443-3319

Future Systems Inc.
12500 W. Cedar Dr.

Lakewood, CO 80228
303/989-0431

Miromit/American Heliothermal Corp.
3515 S. Tamarac Dr./Suite 360
Denver, CO 80237
303/773-6085

R. M. Products
5010 Cook St.
Denver, CO 80216
303/825-0203

Solar Energy Research Corp.
1228 15th St.
Denver, CO 80202
303/573-5499

Solaron Corp.
300 Gallerina Tower
720 S. Colorado Blvd.
Denver, CO 80222
303/759-0101

CONNECTICUT

Falbel Energy Systems Corp.
P.O. Box 6
Greenwich, CT 06830
203/357-0626

Hubbell
45 Seymour St.
Stratford, CT 06497
203/378-2659

International Environmental
Energy Inc.
275 Windsor St.
Hartford, CT 06120
203/249-5011

Solar Heating Systems Corp.
151 John Downey Dr.
New Britain, CT 06051
203/224-2164

Solar Industries Inc.
100 Captain Neville Dr.

Waterbury, CT 06705
203/753-1195

Spiral Tubing Corp.
544 John Downey Dr.
New Britain, CT 06051
203/244-2409

Sunworks
P.O. Box 1004
New Haven, CT 06508
203/934-6301

Wilson Solar Kinetics Corp.
P.O. Box 17308
West Hartford, CT 06117
203/233-4461

DELAWARE

DuPont Co.
Nemours Bldg./Rm. 24751
Wilmington, DE 19898
302/999-3456

Solar Systems Inc.
323 Country Club Dr.
Rehoboth Beach, DE 19971
302/227-2323

DISTRICT OF COLUMBIA

Business and Technology, Inc.
2800 Upton St., N.W.
Washington, DC 20008
202/362-5591 or
202/244-4902

McCombs Solar Co.
1629 K St., N.W. Suite 520
Washington, DC 20006
202/296-1540

Natural Energy Corp.
1001 Connecticut Ave., N.W.
Washington, DC 20036
202/296-7070

Solartherm
1640 Kalmia Rd., N.W.

Washington, DC 20012
202/882-4000

Thomason Solar Homes Inc.
609 Cedar Rd., S.E.
Washington, DC 20022
301/336-4042

FLORIDA

American Solar Power, Inc.
715 Swann Ave.
Tampa, FL 33606
813/251-6946

Astro Solar Corp.
744 Barnett Dr./Unit No. 6
Lake Worth, FL 33461
305/965-0606

Aztec Solar Co.
P.O. Box 272
Maitland, FL 32751
305/628-5004

Beutels Solar Heating Co.
7161 Northwest 74th St.
Miami, FL 33166
305/885-0122

D. W. Browning Contracting Co.
475 Carswell Ave.
Holly Hill, FL 32017
904/252-1528

Capital Solar Heating Inc.
376 N.W. 25th St.
Miami, FL 33127
305/576-2380

Chemical Processors Inc.
P.O. Box 10636
St. Petersburg, FL 33733
813/822-3689

Consumer Energy Corp.
4234 S.W. 75th Ave.
Miami, FL 33155
305/266-0124

CSI Solar Systems Div.
12400 49th St.
Clearwater, FL 33520
813/577-4228

D & J Sheet Metal Corp.
10055 N.W. 7th Ave.
Miami, FL 33150
305/757-7033

Del Sol Control Corp.
11914 U.S. Highway No. 1
Juno, FL 33408
305/626-6116

Energy Applications Inc.
840 Margie Dr.
Titusville, FL 32780
305/269-4893

Falkner Inc.
6121 Alden Rd./P.O. Box 673
Orlando, FL
305/898-2541

Flagala Corp.
9700 W. Highway 98
Panama City, FL 32401
904/234-6559

Florida Solar Power Inc.
P.O. Box 5846
Tallahassee, FL 32301
904/244-8270

General Energy Devices
1743 Ensley Ave.
Clearwater, FL 33516
813/586-1146

Gulf Thermal Corp.
P.O. Box 13124 Airgate Branch
Sarasota, FL 33580
813/355-9783

Hill Bros. Inc./Thermal Div.
3501 N.W. 60th St.
Miami, FL 33142
305/693-5800

J & R Simmons Construction Co.
2185 Sherwood Dr.
South Daytona, FL 32019
904/677-5832

Largo Solar Systems Inc.
2525 Key Largo La.
Fort Lauderdale, FL 33312
(Mail Only)

Matthews Systems Inc.
P.O. Box 1666
Gainesville, FL 32602
904/376-5222

National Solar Systems
P.O. Box 17348
Tampa, FL 33682
813/933-4382

OEM Products Inc./
Solarmatic Div.
2413 Garden St.
Tampa, FL 33605
813/247-5858

Semco
1091 S.W. 1st Way
Deerfield Beach, FL 33441
305/427-0040

Solar Comfort Inc.
249 S. Grove St.
Venice, FL 33595
813/485-0544

Solar Development Inc.
4180 West Roads Dr.
West Palm Beach, FL 33407
305/842-8935

Solar Dynamics Inc.
P.O. Box 3457
Hialeah, FL 33013
305/921-7911

Solar Electric International
6123 Anno Ave.

Orlando, FL 32809
305/422-8396

Solar Energy Components Inc.
1605 North Cocoa Blvd.
Cocoa, FL 32922
305/632-2880

Solar Energy Products Inc.
1208 N.W. 8th Ave.
Gainesville, FL 32601
904/377-6527

Solar Energy Resources Corp.
10639 S.W. 185 Terr.
Miami, Fl 33157
305/233-0711

Solar Energy Systems
1243 South Florida Ave.
Rockledge, FL 32955
305/632-6251

Solar-Eye Products, Inc.
1300 N.W. McNabe Rd.
Bldg. GNH
Ft. Lauderdale, FL 33307
305/974-2500

Solar Fin Systems
140 S. Dixie Hwy.
St. Augustine, FL 32084
904/824-3522

Solar Heating &
Air Conditioning Systems
13584 49th St. North
Clearwater, FL 33520
813/577-3961

Solar Industries of Florida
P.O. Box 9013
Jacksonville, FL 32208
904/768-4323

Solar Innovations
412 Longfellow Blvd.
Lakeland, FL 33801
813/688-8373

Solar Products Inc./Sun-Tank
614 N.W. 62nd St.
Miami, FL 33150
305/756-7609

Solar Systems by Sundance Corp.
4815 S.W. 75th Ave.
Miami, FL 33101
305/264-1894

Solar Water Heaters of New
Port Richey
540 Palmetto
New Port Richey, FL 33552
813/848-2343

Southern Lighting/
Universal 100 Products
501 Elwell Ave.
Orlando, FL 32803
305/894-8851

Sun Power
10400 S.W. 187th St.
Miami, FL 33157
305/233-2224

Sunseeker Systems Inc.
100 W Kennedy Blvd.
Tampa, FL 33602
813/223-1787

Systems Technology Inc.
P.O. Box 337
Shalimar, FL 32579
904/863-9213

Unit Electric Control Inc./
Sol-Ray Div.
130 Atlantic Dr.
Maitland, FL 32751
305/831-1900

Universal Solar Energy Co.
1802 Madrid Ave.
Lake Worth, FL 33461
305/586-6020

Wilcon Corp.
3310 S.W. Seventh
Ocala, FL 32670
904/732-2550

Wilcox Manufacturing Corp.
P.O. Box 455
Pinellas Park, FL 33565
815/531-7741

W. R. Robbins & Sons
1401 N.W. 20th St.
Miami, FL 33142
305/325-0880

Youngblood Company, Inc.
1085 N.W. 36th St.
Miami, FL 33127
305/635-2501

GEORGIA

Independent Living Inc.
5715 Buford Hwy., N.E.
Doraville, GA 30340
404/455-0927

Scientific-Atlanta Inc.
3845 Pleasantdale Rd.
Atlanta, GA 30340
404/449-2000

Solar Technology Inc.
3927 Oakclif Industrial Ct.
Atlanta, GA 30340
404/449-0900

Southeastern Solar Systems
P.O. Box 44066
Atlanta, GA 30336
404/691–1864

Wallace Company
831 Dorsey St.
Gainsville, GA 30501
404/534-5971

HAWAII

The Solaray Corp
2414 Makiki Heights Dr.
Honolulu, HI 96822
808/533-6464

ILLINOIS

Amcon Inc.
211 W. Willow St.
Carbondale, IL 62901
618/457-3022

A.O. Smith Corp.
P.O. Box 28
Kankakee, IL 60901
815/933-8241

Chamberlain Mfg. Corp.
845 Larch Ave.
Elmhurst, IL 60126
312/279-3600

ITT Fluid Handling Div.
4711 Golf Rd.
Skokie, IL 60076
312/677-4030

Johnson Controls Inc./Penn Div.
2221 Camden Ct.
Oak Brook, IL 60521
312/654-4900

Olin Brass Corp./Roll-Bond Div.
E. Alton, IL 62024
618/258-2000

Pak-Tronics Inc.
4044 N. Rockwell Ave.
Chicago, IL 60618
312/478-8585

Solar Dynamics Corp.
550 Frontage Rd.
Northfield, IL 60093
312/446-5242

Sun Systems Inc.
P.O. Box 155
Eureka, IL 61530
309/685-9728

INDIANA

Solar Energytics, Inc.
P.O. Box 532
Jasper, IN 47546
812/482-1416

IOWA

Lennox Industries Inc.
200 S. 12th Ave.
Marshalltown, IA 50158
414/754-4011

Pleiad Industries, Inc.
RR 1/P.O. Box 57
West Branch, IA 52358
319/643-5650

Solar Aire
P.O. Box 276
North Liberty, IA 52317
319/626-2343

KENTUCKY

Mid-Western Solar Systems
2235 Irvin Cobb Dr.
P.O. Box 2384
Paducah, KY 42001
502/443-6295

MARYLAND

KTA Corp.
12300 Washington Ave.
Rockville, MD 20852
201/568-2066

MASSACHUSETTS

Columbia Solar Energy Div.
55 High St.

Holbrook, MA 02343
617/767-0513

Daystar Corp.
90 Cambridge St.
Burlington, MA 01803
617/272-8460

Diy-Sol Inc.
P.O. Box 614
Marlboro, MA 01752

Kennecott Copper Corp.
128 Spring St.
Lexington, MA 02173
617/862-8268

Sunkeeper
P.O. Box 34
Shawsheen Village Station
Andover, MA 01801
617/470-0555

Sunsav Inc.
9 Mill St.
Lawrence, MA 01840
617/686-8040

Sun Systems Inc.
P.O. Box 347
Milton, MA 02186
617/268-8178

Vaughn Corp./Solargy Systems
386 Elm St.
Salisbury, MA 01950
617/462-6683

MICHIGAN

Dow Chemical, USA
2020 Dow Center
Midland, MI 48640
517/636-3993

Solarator Inc.
16231 W. 14 Mile Rd.
Birmingham, MI 48009
313/642-9377

Solar Research
525 N. Fifth St.
Brighton, MI 48116
313/227-1151

Solartran Co.
Escanaba, MI 49829
906/786-4550

Tranter
735 East Hazel St.
Lansing, MI 48909
517/372-8410

MINNESOTA

A To Z Solar Products
200 E. 26th St.
Minneapolis, MN 55404
612/870-1323

Hoffman Products, Inc.
P.O. Box 975
Willmar, MN 56201
612/235-1400

Honeywell Inc.
2600 Ridgeway Pkwy.
Minneapolis, MN 55413
612/870-5200

Ilse Engineering
7177 Arrowhead Rd.
Duluth, MN 55811
218/729-6858

Sheldahl/Advanced
Products Div.
Northfield, MN 55057
507/645-5633

Solargizer Corp.
220 Mulberry St.
Stillwater, MN 55082
612/439-5734

NEVADA

Richdel Inc.
P.O. Drawer A

Carson City, NV 89701
702/882-6786

S. W. Ener-Tech Inc.
3030 S. Valley View Blvd.
Las Vegas, NV 89102
702/873-1975

NEW HAMPSHIRE

Heliopticon Corp.
P.O. Drawer 330
Plymouth, NH 03264
603/536-1070

Kalwall Corp./
Solar Components Div.
P.O. Box 237
Manchester, NH 03105
603/668-8186

NEW JERSEY

Berry Solar Products
Woodbridge At Main
P.O. Box 327
Edison, NJ 08817
201/549-3800

Calmac Manufacturing Corp.
P.O. Box 710E
Englewood, NJ 07631
201/569-0420

Heilemann Electric
127 Mountain View Rd.
Warren, NJ 07060
201/757-4507

Solar Energy Systems Inc.
One Olney Ave.
Cherry Hill, NJ 08003
609/424-4446

SSP Associates
704 Blue Hill Rd.
River Vale, NJ 07675
201/391-4724

NEW MEXICO

K-Line Corp.
911 Pennsylvania Ave.
Albuquerque, NM 87110
505/268-3379

Sigma Energy Products
1405B San Mateo, N.E.
Albuquerque, NM 87110
505/262-0516

NEW YORK

Advance Cooler
Manufacturing Corp.
Route 146
Bradford Industrial Park
Clifton Park, NY 12065
518/371-2140

Ford Products Corp.
Ford Products Rd.
Valley Cottage, NY 10989
914/358-8282

Grumman Corp./
Energy Sys. Div.
Dept. GR
4175 Veterans Memorial Hwy.
Ronkonkoma, NY 11779
516/575-7062

Hitachi American Ltd.
437 Madison Ave.
New York, NY 10022
212/838-4804

International Environment Corp.
129 Halstead Ave.
Mamaroneck, NY 10543
914/698-8130

Revere Copper & Brass Inc.
P.O. Box 151
Rome, NY 13440
315/338-2401

Solar Energy Systems, Inc.
P.O. Box 625 Sentry Pl.

Scarsdale, NY 10583
914/725-5570

Sol-Therm Corp.
7 West 14th St.
New York, NY 10011
212/691-4632

NORTH CAROLINA

Carolina Solar Equipment Co.
P.O. Box 2068
Salisbury, NC 28144
704/637-1243

Standard Electric Co.
P.O. Box 631
Rocky Mount, NC 27801
919/442-1155

OHIO

Glass-Lined Water Heater Co.
13000 Athens Ave.
Cleveland, OH 44107
216/521-1377

Howard Bell Enterprises Inc.
P.O. Box 413
Valley City, OH 44280
216/483-3249

Libbey Owens Ford/
Technical Center
1701 East Broadway
Toledo, OH 43605
419/247-4355

Mor-Flo Industries Inc.
18450 South Miles Rd.
Cleveland, OH 44128
216/663-7300

Owens Illinois Inc.
P.O. Box 1035
Toledo, OH 43666
419/242-6543

Ranco Inc.
601 W. Fifth Ave.
Columbus, OH 43201
614/294-3511

Solar Energy Products Co.
121 Miller Rd.
Avon Lake, OH 44012
216/933-5000

Solar Heat Corp.
1252 French Ave.
Lakewood, OH 44107
216/228-2993

Solar Home Systems Inc.
12931 West Geauga Trail
Chesterland, OH 44026
216/729-9350

Solar Vak Inc.
P.O. Box 1444
Dayton, OH 45414
513/278-6551

OKLAHOMA

Brown Manufacturing Co.
P.O. Box 14546
Oklahoma City, OK 73114
405/751-1343

Tri-State Solar King, Inc.
P.O. Box 503
Adams, OK 73901
405/253-6562

OREGON

Scientifico Components Co.
35985 Row River Rd.
Cottage Grove, OR 97424

PENNSYLVANIA

Aluminum Co. of America
Alcoa Bldg.
Pittsburgh, PA 15219
412/553-2321

Ametek Inc./
Power Systems Group
1 Spring Ave.
Hatfield, PA 19440
215/822-2971

Enviropane Inc.
348 N. Marshall St.
Lancaster, PA 17602
717/299-3737

General Electric Co.
P.O. Box 8661/Bldg. 7
Philadelphia, PA 19101
215/962-2112

Heliotherm Inc.
Lenni, PA 19052
215/459-9030

Packless Industries Inc.
P.O. Box 310
Mount Wolf, PA 17347
717/266-5673

PPG Industries
One Gateway Center
Pittsburgh, PA 15222
412/434-3555

Practical Solar Heating
209 S. Delaware Dr./Rt. 611
Easton, PA 18042
215/252-6381

Simons Solar Environmental
Systems Inc.
24 Carlisle Pike
Mechanicsburg, PA 17055
717/697-2778

Solar Heat Co.
P.O. Box 110
Greenville, PA 16125
412/588-5650

Solar Shelter
P.O. Box 36
Reading, PA 19603
215/488-7624

Sun Earth Solar Products Corp.
RD 1/P.O. Box 337
Green Lane, PA 18054
215/699-7892

Sunwall Inc.
P.O. Box 9723
Pittsburgh, PA 15229
412/364-5349

RHODE ISLAND

Solar Homes Inc.
2 Narragansett Ave.
Jamestown, RI 02835
401/423-1025

TENNESSEE

ASG Industries
P.O. Box 929
Kingsport, TN 37662
615/245-0211

Energy Converters Inc.
2501 N. Orchard Knob Ave.
Chattanooga, TN 37406
615/624-2608

Oak Ridge Solar
Engineering Inc.
P.O. Box 3016
Oak Ridge, TN 37830
615/482-5290

State Industries Inc.
Cumberland St.
Ashland City, TN 37015
615/792-4371

W.L. Jackson Manufacturing Co.
P.O. Box 11168
Chattanooga, TN 37401
615/867-4700

TEXAS

American Solar King Corp.
6801 New McGregor Hwy.

Waco, TX 76710
817/776-3860

Cole Solar Systems Inc.
440A E. St. Elmo Rd.
Austin, TX 78745
512/444-2565

Northrup Inc.
302 Nichols Dr.
Hutchins, TX 75141
214/225-4291

Solar Systems Inc.
507 W. Elm St.
Tyler, TX 75701
214/592-5343

Soltex Corp.
P.O. Box 55703
Houston, TX 77055
713/780-1733

Solus, Inc.
P.O. Box 35227
Houston, TX 77035
713/772-6416

VERMONT

Sol-R-Tech
P.O. Box G
Hartford, VT 05047
802-295-9342

VIRGINIA

Atlantic Solar Products Inc.
Reston International Center
Suite 227
11800 Sunrise Valley Dr.
Reston, VA 22091
703/620-2300

Helios Corp.
1313 Belleview Ave.
Charlottesville, VA 22901
804/293-9574

Reynolds Metals Co.
P.O. Box 27003
Richmond, VA 23261
804/281-3026

Solar American
106 Sherwood Dr.
Williamsburg, VA 23185
804/229-0657

Solar Corp. of America/
Intertechnology Corp.
100 Main St.
Warrenton, VA 22186
703/347-7900

Solar One Ltd.
709 Birdneck Rd.
Virginia Beach, VA 23451
804/422-3262

Solar Sensor System
4220 Berritt St.
Fairfax, VA 22030
703/273-2683

WASHINGTON

Ecotope Group
747 16th Street, E.
Seattle, WA 98112
206/322-3753

E&K Service Co.
16824 74th Ave., N.E.
Bothell, WA 98011
206/486-6660

WISCONSIN

Solaray Inc.
324 S. Kidd St.
Whitewater, WI 53190
414/473-2525

Sun Stone
P.O. Box 941
Sheboygan, WI 53081
414/452-8194

Annual Energy Requirements of Electric Household Appliances*

The estimated annual kilowatt-hour consumption of the electric appliances listed in this reference is based on average-size appliances and normal use.

	AVERAGE WATTAGE	EST. KWH CONSUMED ANNUALLY		AVERAGE WATTAGE	EST. KWH CONSUMED ANNUALLY
HEALTH & BEAUTY			Sandwich Grill	1,161	33
Germicidal Lamp	20	141	Toaster	1,146	39
Hair Dryer	381	14	Trash Compactor	400	50
Heat Lamp (infrared)	250	13	Waffle Iron	1,200	20
Shaver	15	0.5	Waste Dispenser	445	7
Sun Lamp	279	16			
Tooth Brush	1.1	1.0	FOOD PRESERVATION		
Vibrator	40	2	Freezer (15–21 cu. ft.)		
			chest type, manual defrost	—	1,320
HOME ENTERTAINMENT			upright type		
Radio	71	86	manual defrost	—	1,320
Radio/Record Player	109	109	automatic defrost	—	1,985
Television			Refrigerators/Freezers		
black & white			manual defrost,		
tube type	100	220	10–15 cu. ft.	—	700
solid state	45	100	automatic defrost,		
color			16–18 cu. ft.	—	1,795
tube type	240	528	automatic defrost,		
solid state	145	320	20 cu. ft. & up	—	1,895
HOUSEWARES			LAUNDRY		
Clock	2	17	Clothes Dryer	4,856	993
Floor Polisher	305	15	Iron (hand)	1,100	60
Sewing Machine	75	11	Washing Machine		
Vacuum Cleaner	630	46	(automatic)	512	103
			Washing Machine		
FOOD PREPARATION			(nonautomatic)	266	76
Blender	300	1	Water Heater	2,475	4,219
Broiler	1,140	85	(quick-recovery)	4,474	4,811
Carving Knife	92	8			
Coffee Maker	894	106	COMFORT CONDITIONING		
Deep Fryer	1,448	83	Air Cleaner	50	216
Dishwasher	1,201	363	Air Conditioner (room)	860	860†
Egg Cooker	516	14	Bed Covering	177	147
Frying Pan	1,196	100	Dehumidifier	257	377
Hot Plate	1,200	90	Fan (attic)	370	291
Mixer	127	2	Fan (circulating)	88	43
Oven, microwave (only)	1,450	190	Fan (roll-away)	171	138
Range			Fan (window)	200	170
with oven	12,200	1,175	Heater (portable)	1,322	176
with self-cleaning oven	12,200	1,205	Heating Pad	65	10
Roaster	1,333	60	Humidifier	177	163

*Based on 1,000 hours of operation per year. This figure will vary widely depending on area and specific size of unit.

†Figures developed by Edison Electric Institute, 90 Park Ave., New York, NY 10016.

BIBLIOGRAPHY

Energy Activities With Energy Ant. Federal Energy Administration. 1974. 28 pp.
    Available from Superintendent of Documents, U.S. Government Printing Of-
    fice, Washington, DC 20402. Order number 041–018–00542; $1.40.*
    A general overview of energy and conservation in a coloring book/workbook
    format. Looks good for third to sixth graders.

The Fuel Savers—A Kit of Solar Ideas for Existing Homes. Dan Scully, Don Prowler,
    Bruce Anderson. Total Environmental Action, Harrisville, NH 03450. 1976. (pa-
    perback, 8½ x 11, 59 pp.)
    An excellent guide for the handyman, do-it-yourselfer, and small contractor.

Measures for Reducing Energy Consumption for Homeowners and Renters. Office of
    Energy Systems, Federal Power Commission, Washington, DC 20426. March
    19, 1975. Available from National Technical Information Service, Springfield,
    VA 22161; order number PB 240–472.

Slash Energy Bills. Allen Lawrence, P.O. Box 13622, San Antonio, TX 78213. 1976.
    $1.50,* (paperbound, 8½ x 11, 28 pp.)
    A complete and readable guide containing many useful thoughts, hints, and
    energy consumption tables.

Solar Hot Water and Your Home. A National Solar Heating and Cooling Information
    Center Publication. First distributed August 1977.
    The National Solar Heating and Cooling Information Center maintains extensive
    files on all aspects of solar hot water heating. Additional information may be ob-
    tained by writing to Solar Heating, P.O. Box 1607, Rockville, MD 20850.

Tips for Energy Savers. Available from Consumer Information, Public Documents
    Center, Pueblo, CO 81009.
    Loaded with useful tips, this pamphlet can be obtained free of charge from the
    address noted above.

*Please contact source for latest price before ordering as prices subject to change.

6

Our Poll—Solar Homeowners*

We sent detailed questionnaires to the owners of all the solar homes in America listed as "completed" in William A. Shurcliff's *Solar-Heated Buildings—A Brief Survey* (9th edition). We wanted those whose systems had been in use through at least one winter. Over one hundred homeowners were involved; eighty-eight responded. This unusually high number of responses confirmed our belief that solar homeowners would want to share their experiences in the hope of correcting some of the problems they'd encountered. The responses were generous, at times even passionate. Here's what we found:

SUMMARY

58% did their own solar installations.

27% were fairly well satisfied with their systems, although more listed things they'd do differently if given a second chance.

24% had enough trouble with either their solar installation contractors, or their subsequent servicing contractors, to give their trouble more than passing reference.

21% had trouble with leaks.

17% had problems with their heat-storage facilities.

12% felt they'd been sold unsuitable or defective equipment.

*Not to be confused with the survey of solar heat *feasibility* for *existing* homes, published under the title *Your Home's Solar Potential*, by the authors of this book (64 pages, $9.95, Edmund Scientific Company, Barrington, NJ 08007). For more on this subject, see chapter 4.

6% were very unhappy with their solar systems, and an equal number,

6% were totally delighted with their solar systems.

On reading these results, you might tend to assume that your solar experience is likely to be no better than those summarized above, but that's not the case. First of all, there is a tendency, in responding to any survey, to overdo the complaints. It's a good outlet for any pent-up frustrations experienced in connection with the subject under discussion. We all like to moan a bit. It's due, in part, to the fun of complaining, and in part to the fact that we've gotten so far away from doing things for ourselves we almost ask to be taken. Consider the auto repair industry, for instance. We've become a nation of pigeons.

Well, here's our chance to get even. Solar heat is simple, understandable. If you use this book carefully you can buy your system and its service without the threat of so many things going wrong. You may even want to design and build your own system, using some of the many good books now available on that subject (using this one as your trouble-shooting guide). Notice that more than half the respondents to our poll installed their own systems. They were the people who voiced the fewest complaints. Some of that may have been due to pride, but it seems fair to assume, too, that one will do a far more careful job for himself than for others. Undoubtedly, however, the do-it-yourself portion of the solar market is going to shrink. With more and more manufacturers and contractors getting into the business, and with the big consumer push now under way, the owner-installer share of the market may well drop from 58% to less than 20%. But so what? The important things are to keep your system simple, to keep control of your own sunlight, and to understand enough of what is going on to avoid being taken, either during, or in the years after, installation.

The bugs are being worked out of what was formerly troublesome equipment, and the emerging solar industry—design, manufacturing, installation, and service—is building up a fund of experience that will help reduce the number of problems you'll face.

PUZZLER

Of all the thousands of words we received in response to our poll, none were more puzzling than these: "tar on rug, shaved off." How the tar got there in the first place, who shaved it off, and what it is that tar

The Junius Eddy home in Little Compton, Rhode Island 02837. This home uses all three basic solar systems; active-air, active-water, and passive. The small structure to the right (three 7 x 3 ft. collectors) holds the active system for heating domestic water, while the space heating is done by the 18 (the two at far left are windows) active-air collectors. Below are the passive windows with their insulated shutters.

The Bob and Nancy Homan house "Solaria," in Indianmills, New Jersey. Designed by Malcolm B. Wells, coauthor of this book, the building uses a Thomason, trickle-type active solar heating system.

Earth-sheltered northern exposure.

South-facing Thomason trickle-type collectors.

has to do with a solar heating installation, were all left to our imaginations, and we leave them to yours.

MISADVENTURES

Like any project, the installation of solar heating can be a great adventure—*if* you're willing to roll with the punches a bit. After all, *something* is bound to go wrong on every project, so you'll be a lot better equipped to deal with it if you're able to see its funny side. Listen to these experiences:

"I fell 42 feet while working on my rooftop collector, and walked away from the incident smiling, if a bit shaken. Luckily, I landed on my feet—in some freshly plowed soil. Looking back on the experience, I'd have to say that solar heat provided one of the biggest thrills of my life."

Dumb questions seem tailor-made for solar heating, according to many owners. "Can you get a shock from it?" "Are the rays dangerous?" "Is that thing really supposed to work?" "What good is heat, up on the roof?" If you stand there and explain all the principles of solar energy to everyone who asks about them you'll never get anything else done. The best defense against a dumb question is a dumb answer.

Sundown House at Sea Ranch, California, designed by David Wright A.I.A., environmental architect. Notice the solar collection area on the wall surface of the raised portion of the structure, on the left of the photo.

Try some of these:

• "Can you get a shock from it?" Not if your television set is turned off.

• "Are the rays dangerous?" Not particularly, but I wouldn't want to stand near it if I were pregnant.

• "Is that thing really supposed to work?" No, we're just doing it to impress the neighbors.

• "What good is heat, up on the roof?" We moved all our furniture into the attic, and live up there all winter.

"1,000 plastic jugs full of water can get mighty heavy if you have to heave them, from the back of a pickup truck, up to a second-story window. They store heat beautifully, but, at 8 pounds apiece, you soon begin to realize that throwing a thousand of them 10 feet in the air is the work-equivalent of climbing to the top of the Washington Monument."

The Henry Mathew home in Coos Bay, Oregon. This active solar heating system uses untreated water in ¹/₂ in. galvanized iron pipe, wired snugly to a sheet of aluminum and painted black, with a single glazing of ¹/₁₆ in. glass. The south portion of the roof is covered with aluminum foil, increasing the efficiency of the collectors by about 50 percent.

The Charles Marsh home in Pomona, New Jersey. The 4 x 16 ft. aluminized Mylar reflector below the southeast-facing wall-mounted collectors increases solar heat gain to better than 700 BTU/ft²/day.

"They said they'd kill me if I mentioned my fuel bills in the company cafeteria. Everyone else was using gas or oil or electricity at a phenomenal rate last winter, while I was spending $23 a month. They invited me in no uncertain terms not to say a word."

"Goats can do a lot of damage to a solar collector. My house is set so low into the ground for weather protection the goats can reach the roof. The earth-shelter is great but all our collector covers got nibbled until I rigged a fence to keep the goats away."

LINES TO REMEMBER

This then-still-unwritten book got some nice endorsements from our poll respondents. They made us more determined than ever to help make your solar adventure a pleasant one. . . .

"I hope you're serious about your book because it really is necessary."

"Please help others avoid these troubles (poor design and faulty installation)."

"Because of my engineering training I was able to check the design and resize all the ducts, but this makes me wonder what an average person is to do if he wants solar heat and doesn't know how to design or install it."

PROBLEMS

The two problems almost never mentioned in our poll results, of course, are the biggest ones of all, and, even though they're covered in detail in other chapters, it's important that we stop here for a moment to reemphasize them:

1. *Inadequate sunlight,* and
2. *Inadequate home insulation.*

They were seldom mentioned by our respondents because most of them did their homework in that regard. It's the *future* solarizers, those who may not have the pioneering spirit of their recent predecessors, who may need reminding.

You'd think that those two problems would be at the top of every potential solar homeowner's priorities list, but again and again we deal with people who hope to bend the rules, to cheat nature, and that is one place in which cheating is never successful. When you're dealing with natural forces you're dealing with laws that can't be broken. There's no pushing up the solar thermostat if you've disregarded the fundamentals. Your only alternatives in that case are to go on using expensive and wasteful fuels—or freeze.

Now, with that understood, we can turn our attention to the prob-

The William H. Knepshield Home in Mifflinburg, Pennsylvania. Passive heat is supplied by the 178 ft.² of double-glazed windows on the south wall. At the far left you can see a portion of the 250 ft.² air-active heating system.

lems encountered by all the people who responded to our national survey—people with actual solar heating experience. Continuing the numbering system started with the Big Two, above, we find this to be problem number—

> 3. *High costs.* Listen: "I blame it all on inexperienced estimators." "Our cost doubled, and we couldn't do anything about it." "The contractor was learning at our expense, paying high-priced plumbers." "Our $10,000 solar system ended up costing $16,000."

We'll deal with the ways of handling this, and all the other problems, in other parts of this book, so don't be disappointed if we don't tell you, right here, how to hold costs in line. This poll summary would get hopelessly bogged down if we tried.

These are the other problems that turned up, listed in the apparent order of their importance to our respondents.

> 4. *Delays.* "They never delivered anything on time. They'd promise and we'd stay home and they wouldn't show up." "Failure to man the job. Lost weeks of good construction weather."

> 5. *Water leaks* (or leaks of antifreeze-treated water). "Hoses leaked at panel couplings on roof." "Ice formed between panels because covers weren't installed over joints. Ice pressure broke fittings." "They've got to solve the expansion problems. The collectors can go from 0° in the morning to boiling by noon." "The collector-cover caulking got hard and leaked." "Trickle systems are notorious leakers." "Plastic gallon jugs [for heat storage in air systems] will leak if stacked. You've got to stand them on individual shelves."

> 6. *Air leaks.* "A lot of the heat in our storage pit escaped at night because of dampers that wouldn't close tightly." "The industry standard for damper-tightness is only one-tenth as strict as it should be." "Our [Trombe-type, passive, heat-storing concrete] wall tends to create reverse thermosyphoning at night, filling the lower floor with cold air." "A lot of air leaks out through the duct joints." "Whenever we go away, all the heat leaks out of the [rock-filled, air-] storage bin, and

we have to use all the solar heat we get for the next couple of days, heating up the rocks again."

7. *Design flaws.* "They didn't test the water, so we get rust-colored streaks on our [trickle-type, open-under-glass, black] panels." "Watch out for corrosion! Mixing untreated hot water with aluminum and copper is asking for trouble." "Don't use a completely sealed system; there's no way to add antifreeze or drain the pipes." "Arrange the system so it drains into a reserve tank when it's not in use; don't let all those pipes sit there all summer, full of water." "Keep it simple, understandable. It's terribly important that we not be overwhelmed by the look of what is basically a very straightforward technology." "They told the contractors to mount our panels at the wrong angle, so we wasted a lot of sunlight." "Use more reflectors. You can reduce expensive collector area with them." "Don't have too much airspace in the collector." "Provide for the release of steam if boiling temperatures are

The Lawrence H. Carr home eight miles south of San Luis Obispo, California. At the left of the photo is the owner/designer, Lawrence Carr, showing the coils of ¹/₂ in.-diameter black polyethylene tubing, which face the two 20 x 20 ft. collectors. During winter months, both collectors are covered with a single glazing of clear (.006 in.) polyethylene.

The Doug and Meg Kelbaugh home in Princeton, New Jersey. Doug is the architect who designed this passive solar heated home, using an adaptation of Professor Trombe's work in the French Pyrenees.

The Daniel Zwillinger home in New Boston, New Hampshire. Mr. Zwillinger has a patent pending on this air system which he designed and claims is quite revolutionary in concept, quality, performance, and cost.

possible in your collector." "Our air ducts were undersized by half, putting a big load on the fan." "Don't use manual dampers. You get tired quickly of turning half-a-dozen handles every day to regulate the system.""Keep the house-heating and domestic hot water systems separate so you can service one without shutting down the other."

8. *Controls.* "Our controls are overdesigned. The utility company put them in for us." "Nobody will service our controls, they're so complicated." "Label all the controls clearly, then keep a notebook near them to record what you did and how it worked." "It takes a long time to debug the system."

9. *Equipment and material failures.* "Fiberglass covers are hard to get smooth and tight on a collector. They buckle." "The fans are too noisy."

10. *Workmanship (installation and service).* "It's a crime the way they put this stuff together." "We have to call for days to get service." "The builder never followed the design drawings."

11. *Inexperienced contractors.* "The plumber had to ask *me* where the main pipe was supposed to connect!"

12. *Frustrations in dealing with government agencies.* "ERDA [The Federal Energy Research and Development Administration] is impossible." "Be prepared to wait for months."

13. *Miscellaneous.* "There aren't enough contractors doing solar heat work." "Make sure your indoor (insulating) shutters are easy for women and children to operate. In mild climates like the Southwest, 'space blankets' (mirrored Mylar film) are all you need to cover your windows at night."

That's it. That's the worst of what's happened to America's solar homeowners. We asked them, "What would you do differently now, if you had the opportunity?"

Some of the advice is woven through the chapters you've completed, but much of it relates to their contracting, legal, and financing experiences. So, read on and we'll see if we can tell you, now, how to make your experience one of the better ones.

APPENDIX/CHAPTER 6

Solar Questionnaire

The following is a copy of the original one-page questionnaire sent to most of the solar homeowners in the United States. Over one hundred homeowners were polled, and eighty-eight responded.

SOLAR QUESTIONNAIRE FOR THE _____ SOLAR BUILDING

	Yes	No

1. Have you been through at least one winter with your solar building? . . . .   ____ ____

2. If yes, for how many years has your solar system been installed and functioning? . . . . ____ years.

3. Does your system provide enough heat for your comfort? . .   ____ ____

4. Did you do your own solar hardware installation? . . . .   ____ ____

5. If a contractor was used for the installation, what was your greatest problem (frustration) in dealing with him? _____
   _____
   _____
   _____

6. Do you perform maintenance on your own solar system? . .   ____ ____

7. If maintenance is purchased, what is your greatest problem (frustration) in obtaining this service?_____
   _____
   _____
   _____

8. Would you consent to a brief telephone interview on this subject (to be used strictly for our new book)? . . . .   ____ ____

   If yes, at what number can you be reached during weekday evening hours? ( ) ____-____

9. Do you recall any humorous experiences relating to your installation or use of solar heating?

10. What do you consider the greatest problem with your installation to be, and what would you do differently now, if you had the opportunity?

(Please use the back of this questionnaire if additional space is needed.)

Solar Buildings in the United States

Most of the homeowners polled are included in the following list, taken from *Solar-Heated Buildings—A Brief Survey,* 9th edition (available from William A. Shurcliff, 19 Appleton Street, Cambridge, MA 02138).

ALABAMA
Huntsville
  MSFC Solar Test House

ARIZONA
Amado
  Bliss House
Phoenix
  AFASE House
  Prototype Skytherm H's.
Prescott
  Prescott House
Tucson
  Decade 80 Solar House
  Meinel House
  U. of Ariz. Env. Res. H's.
  U. of Ariz. Solar Lab.
Yuma
  Hansberger Apartments

CALIFORNIA
Atascadero
  Skytherm SW House (Hay)
Davis
  Ward House
Death Valley
  Scotty's Castle
Del Mar
  Beckstrand House
El Cajon
  En. Systems Inc. House
Kentf'd Hills
  Jeffrey House
Los Altos
  Breuch House
San L. Obispo
  Carr House
Santa Clara
  Community Recr. Center

Valley Center
  Mills House
Visalia
  Cottrell House

COLORADO
Beulah
  Sunray II (Fredregill)
Boulder
  Boulder House
  Bushnell House
  W. Pearl Cond., S. Bldg.
  W. Pearl Cond., N. Bldg.
Colo. Springs
  Jackson House
  Phoenix Col. Sp. H's.
  Wood House
Denver
  Crowther H's, 419 St. P.
  Crowther H's, 435 St. P.
  Crowther Retrofit H's.
  Crowther Solar Of. Bldg.
  Gump Glass Co. Bldg.
  Löf House in Denver
Fort Collins
  Colo. St. U. Sol. H's. I
  Colo. St. U. Sol. H's. II
  Colo. St. U. Sol. H's. III
Livermore
  Shippee House
Pueblo
  Sunray I House
Snowmass
  Shore House
Wellington
  Perna House
Westminster
  Com. Coll. Denv. N. Campus

CONNECTICUT

Guilford
  Barber House
  Pinchot Guest House
Hamden
  Housing for Aged in Conn.
New Milford
  Lasar House
Stamford
  Pyramidal Optics House
Westbrook
  Shoreline H's. (Sol-a-Sea)

DELAWARE

Newark
  Solar One House

FLORIDA

Gainesville
  Univ. of Fla. Solar H's.

GEORGIA

Atlanta
  Towns Elementary School

ILLINOIS

Carlyle
  Jantzen House
Eureka
  Eureka Solar House

MAINE

Bar Harbor
  R. & N. Davis House
Orono
  Hill House

MARYLAND

Calvert County
  Thomason Solar House 6
Gaithersburg
  NBS Solar House
Pr. G. County
  Thomason Solar House 1
  Thomason Solar House 2
  Thomason Solar House 3

  Thomason Solar House 4
  Thomason Solar House 5
Timonium
  Timonium School

MASSACHUSETTS

Acton
  Acorn Structures House
Boston
  Grover Cleveland School
Cambridge
  MIT Solar House I
  MIT Solar House II
  MIT Solar House III
Concord
  Thornton House
Dover
  Peabody House
Lexington
  MIT Solar House IV
Lincoln
  Gras House
Manchester
  Merchant House
Marlboro
  DIY-Sol, Inc., House
No. Chelmsford
  N.E. Tel. & Tel. Bldg.
Waltham
  Solar Heat Test House
  Solar Heat House #2
Weston
  Saunders House

MICHIGAN

Saginaw
  Federal Bldg.

MINNESOTA

Brooklyn Park
  North View J. Hi. Sch.
Minneapolis
  Honeywell Mobile Lab.

Rosemount
  Ouroboros-South House
St. Paul
  Ouroboros-East House

NEBRASKA
Mead
  Solar, Inc., Of. Bldg.

NEVADA
Boulder City
  Desert Res. Inst. Lab. Bldg.

NEW HAMPSHIRE
Bedford
  Tyrrell House
Etna
  Sol-R-Tech Solar House 1
Manchester
  Federal Office Bldg.
Marlow
  Marlow House
New Boston
  Zwillinger House

NEW JERSEY
Princeton
  Princeton Solar Lab.
Shamong Twp.
  Solaria (Homan House)

NEW MEXICO
Albuquerque
  Bridgers-Paxton Of. Bldg.
Corrales
  Zomeworks House (Baer)
  P. Davis House
Las Cruces
  N.M. Dept. Agric. Bldg.
Los Alamos
  Nat'l. Sec. & Res. Bldg.
Santa Fe
  Allers House
  van Dresser House
  Seton Village House

Terry House
  Wright House
State College
  State College Solar H's.

NEW YORK
Hopewell Jcn.
  Ramage House
Millbrook
  Cary Arboretum Bldg.
New Paltz
  Eccli House

NORTH CAROLINA
Fairview
  Woodward House

OHIO
Columbus
  Ohio St. U. Solar House
Valley City
  Howard Bell Ent. House

OKLAHOMA
Wagoner
  Engle House

OREGON
Bend
  Young House
Coos Bay
  Mathew House
Eugene
  Eugene House
Roseburg
  Erwin House

PENNSYLVANIA
Harmarsville
  Learning Model Of. Bldg.
Mifflinburg
  Knepshield House
Mount Holly
  Slyder House
Schnecksville
  PP&L Energy Cons. House

Stoverstown
  Lefever House
Valley Forge
  G. E. Mobile Home
  G. E. Solar Heated Bldg.

RHODE ISLAND
Jamestown
  Solar Homes Prototype
So. Kingstown
  Carpenter House

SOUTH CAROLINA
Cheraw
  Technical College Lab.

VERMONT
Marlboro
  Mumford Cottage #2
Norwich
  Norwich House

Warren
  Dimetrodon Condominium
White R. Jcn.
  Custom Leather Boutique
Windham
  People/Space Co. House

VIRGINIA
Fairfax County
  Miller & Smith Office
Greenway
  Madeira Science Bldg.
Hampton
  NASA Eng'g. Bldg.
Richmond
  Phys. Sci. Museum HQ
Warrenton
  Fauquier County Hi. Sch.

WYOMING
Lovell
  Bighorn Canyon Center

7

Contracts and Contractors
(and Architects)

Where do you live, anyway? Michigan? California? Massachusetts? Georgia? Missouri? Oregon? The fast-food joints outside your town may look exactly like the ones outside ours, but, as a people, we aren't nearly as identical as a casual visitor to this nation might assume. Thank God. The endless richness and diversity of this wide country are both refreshing and encouraging, but they're a real problem for anyone who tries to write about a subject as complicated and important as this one. You'd need a small library to cover the subjects of contracts and contractors, state by state, in any kind of detail, so we're not even going to try. Instead, we're going to offer a big, safe, generality and then develop it as far as we think best.

This entire chapter, in fact, can be summed up in a single sentence: In order to have a successful relationship with a contractor you must—

1. check him out, and
2. make him bid competitively; then
3. enter the contract in good faith, remembering this:
4. don't overpay him during construction, and
5. keep a written record of everything.

There's a lot more, of course: insurance coverage, guarantees, contract terms—all that business, some of which we'll get to as we go along, but if you follow the five basic rules you'll probably have little trouble.

We almost cried when we first read the comments some of our national solar homeowners' poll respondents sent us. "The cost went from $10,000 to $16,000 and there was nothing we could do." Or: "He took all our money and hasn't come back to finish the work."

In this real-life world, a certain amount of trouble seems to be inevitable at times, almost necessary. No one can prevent every accident and mistake. But when we see how unnecessary some of our respondents' troubles were we yearn to roll back the calendar and let those unhappy people read this book before they start over again. In solar contracts, as in everything else, a bit of good advice can eliminate a lot of grief.

Let's take our big five-part rule for dealing with contractors and expand it step by step:

STEP 1: CHECK HIM OUT

In spite of all the negative things we all hear about the construction industry and its notorious little brother, the home improvement business, we've found most contractors to be fairly honest, reasonably dependable, and interested in doing a good job. They're also interested, of course, in making money. Like the rest of us, they're still a bit spoiled from having had a twenty-five-year free ride between 1950 and 1975, but, now that the great building boom has subsided, they're learning to give better value for each dollar.

Occasionally, however, you will find a contractor who is an absolute louse; someone who wants to do as little of your work and take as much of your money as he possibly can. If you get bad vibes from a person on first meeting him, be doubly careful. You weren't given your vibe-antennas for nothing. Sometimes, of course, you won't recognize a con man at first sight. That's why we added those rules about staying a little behind in your payments and keeping good records. Usually, however, all the warning signs are there before you sign on the dotted line—*if* you're willing to recognize them.

Every reputable contractor will give you the names of his most recent customers. But if *you* fail to follow through and only pretend to check his references you'll have no one to blame but yourself if things go wrong. Check not only his recent customers but also his credit rating. Your bank can tell you how to do this. It's important.

It really pays to find out what the recent customers of your contractor have to say. In order to hear what you need to hear, you'll need a sensitive inner ear. Be prepared, for instance, to hear *some* complaints about the guy, no matter what he's really like. You have to recognize comments like "The damned thing works all right but he still hasn't taken all his cartons away" (even when such comments are delivered in the most outraged of voices) as being of a far different nature from this: "He never paid any of his suppliers and he hasn't come back for two months." All of us have weaknesses and strengths; nobody's perfect. Try to get a sense of how the contractor performs on the important items, and be flexible enough to live with some temporary frustrations on the others. After all, he may have some problems with you, too.

STEP 2: MAKE HIM COMPETE

Unless the contractor is your dearest friend—or your own brother—we think it's always wise to make him compete with others for the chance to do your work. In fact, even if he *is* your dearest friend it's still not a bad idea. If your job is too small to attract competition, of course, you may have to take anyone who is available, but if your job is *that* small why aren't you doing it yourself?

Competition is the backbone of the free enterprise system. When there is real competition the consumer is protected. When there is collusion—when the bidding is rigged—there is only charade. In order to be treated fairly by your bidders you must start by being fair yourself; *don't ask—or allow—anyone to bid unless he's someone to whom you'd be willing to award the job if his price were right.* Contractors spend a lot of money preparing bids, particularly those which involve alterations to existing structures, and it is not right to ask for bids if your only motive is to sharpen the price of another contractor, one to whom you've already planned to award the contract. You may get away with it, but if the quality of human society continues to deteriorate you can no longer have the fun of blaming others.

One more point on the morality of bidding: award the job to the low bidder. Unless your project is to be paid for with public funds you are under no legal obligation to do this, or even to divulge to any bidder the bids of his competitors. You can award the job to anyone. But you'll help each contractor and all future customers by sharing the bid results. It's wrong to negotiate with a favorite among the bidders, using the lower prices of others as your lever. Unless unusual conditions prevent it, the right thing to do is award the job to the lowest bidder, or throw out all the bids and start over.

On small construction contracts, bids are often called estimates, but an estimate is literally "a rough calculation," and rough calculations have a tendency to escalate when they're turned into smooth ones, so get written bids (firm prices) and try to make sure that all the bidders are bidding on identical, or at least comparable, systems. If one is bidding on Solarapex and another on Solaracme you may not be able to make a wise decision. If each bidder is tied to a different solar supplier, then don't name *any* manufacturer or supplier when you ask for bids; write what we call a *performance specification*, describing the kinds of materials and results you want, forcing each bidder to meet your specifications with his bid. Sometimes performance specs will name a manufacturer as an example of a company providing the kind of material or equipment you want; the name is, in that case, followed by the words "or approved equal." This is somewhat loose since there can be disputes over what is truly equal to what, but at least it's a guide to a general level of acceptability.

Should you ask your solar contractor to install your home insulation? Our feeling is that you should not. Having read chapter 3, you know what you need, and can either do it yourself or have it done by a handyman or insulation contractor without paying the extra cost of having such work supervised by the solar contractor.

There's a whole world of potential legal problems in a construction contract, but then there are just as many in the operation of a motor vehicle, and in marriage, and in everything else. You shouldn't be put off by the threat of problems. You should instead act with greater confidence for having been made aware of them. The *easiest* procedure, of course, if you can afford it, is to hire professionals for every step of the process . . . a solar specialist/mechanical engineer to design your system and administer the construction of it, and a lawyer to handle the contracts and other legal matters. Where a lot of money or a lot of complications are involved, that is the only procedure to follow, but on most home solar installations it is not necessary. The American economy would grind to a halt if we all tried to do it that way. There just aren't enough engineers and attorneys to handle every little construction contract, anyway. If you're careful—and fair—you should be able to buy your new solar system without, as we say, getting burnt.

In the language of construction, one good word to remember is "provide." Make it clear that you want the contractor to do more than simply furnish—or install—the work. You might even write that into the contract—*that he is to provide a complete, functioning, fully guaranteed system*. A nice, big general catchall like that may get you into less trouble than will a long and detailed spec to which you may have failed to add a crucial clause.

Construction guarantees usually run for a year after final payment, or after completion of the work. Often, the individual components and equipment will carry longer guarantees or warranties. We'll get deeper into that when we discuss contracts. Here, with regard to competitive bidding, it's only necessary to remind you that the bidders must be told that among your requirements for the system is one stating that the solar contractor himself must guarantee all workmanship and materials for a full year.

"But what if I don't know how to describe the system I want? How can I get competitive bids?" We often hear that kind of question and we always answer it the same way: if you can afford professional help, get it. Sometimes, an hour or two of a mechanical engineer's, or a solar consultant's, time is all you'll need. Ask him first what his hourly rate is and then take advantage of it. But if you can't afford such services, or if, as it is to most people, the land of lawyers and consultants is a foreign one, then your best bet is to educate yourself, to read a book like this one—and others, too, if you can handle the technical side of solar devices—and then write a brief description of the system you want. Here's an example:

John Butz, President
Solaracme Contracting Company
18 Main Street
Middleville, IL

Re: Request for a Bid on a
Solar Heating System for the house of
Elizabeth Jones
421 Center Street
Middleville, Illinois

Dear Mr. Butz:

During the winter of 1977–1978 my total heating bill was $827.20. I believe my house is suited for solar heating. If you are interested in providing it with a solar system will you please send me a bid for the work? I must have your proposal no later than September 21, 1978. Here's what I want:

> An air system with fully watertight and airtight rooftop collectors of at least 400 square feet, forming with the roof a permanently

weatherproof structure, and having an insulated rock bed storage unit in the basement, complete with all ducts, controls, parts, and connections to my present heating system as needed to reduce my heating bill (for a winter equivalent to that of 1977–1978) by 60%. It is important to me that all parts of the system be selected for low maintenance, resistance to corrosion and leakage, ease of operation, and high efficiency.

In addition, I want you to provide a separate, liquid-type solar preheating system for my domestic water heating system, using copper for all parts having contact with water or the collector liquid. Include provisions for easy filling of the system, and a reserve tank into which it can be drained. This system must have its own collectors of at least 35 square ft. and a thoroughly insulated solar-heated domestic water storage tank of 80-gallon capacity, providing all work needed for a complete system.

Please name in your proposal the manufacturers of all major components.

I look forward to your response on September 21.

Sincerely,
Elizabeth Jones

That's only one of an infinite number of different combinations you might describe; by using it as a skeleton, and chapters 2, 3, and 4 as your guides, you should be able to improvise your own quite successfully.

If you wonder why such a letter asked for a solar heating system to reduce the heating bill by only 60%, remember that unless that percentage stays somewhere near the percentage of performance we predicted in chapter 2 for your area, the cost will tend to get out of hand and the efficiency of the system may drop. Once you've got your 60% solar system there's plenty of time to tune it up to a higher level by using more home insulation, more reflector area, lower thermostat settings, and perhaps even the soon-to-appear microcomputers for the monitoring and control of home energy systems.

Obviously, if you live in New Mexico, your system will be far, far different from the one you'd use in Maine, but the bid request for the construction of each system, can, at least in spirit and tone, be identical. You have to strike a careful balance between scaring the contractor away with overly technical and legalistic requirements, and inviting him to take you to the cleaners with overly loose ones.

OK. You've sent your bid-request letter to each of—say—four

recommended solar contractors* and they have submitted their proposals (bids). Now what? Now you have to read the proposals carefully and see if—

> 1. there is among them at least one which meets your budget, and
> 2. it includes everything you asked for (otherwise the bids will not have been competitive).

One good way to get added protection is to have your bid-request letter included by reference in the proposals. But that's easier said than done.

It's a good idea, then, to meet with the lowest bidder whose bid covered everything you asked for, discuss the whole project with him (keeping notes of the conversation), and make arrangements for the signing of the contract. Contractors' proposals are often presented on forms which, when signed, become contracts. Just be careful to see that the terms are fair. More often than not, such care will result in the writing of a new proposal, inasmuch as your meeting with the contractor, and your subsequent review of his proposal, will undoubtedly result in new understandings and modifications which will require a rewriting of the original submission.

STEP 3: ENTER THE CONTRACT IN GOOD FAITH

When it comes to the question of contract terms, the best advice is always to get the help of a lawyer. The material we offer is not professional legal opinion, so you must take it more as a bit of advice from a friend.

We thought that a good way to cover this subject (construction contracts) would be to print a couple of actual agreement forms, filled in with the terms of a fictitious agreement, and then flag some points we thought worth mentioning. This first sample is on a form† developed and refined over the years by the American Institute of Architects. It's designed to be fair, favoring neither the owner (you) nor the contractor, and it's been tested in many, many courtrooms. The only trouble here is it's written as if an architect were involved. But take a look anyway.

---

*During the last years of the seventies you may find mostly plumbing and heating contractors doing this work, and you may even have to educate them a bit; but it appears to us that by the early eighties special solar firms will be doing most of the installation and maintenance work.

†Reprinted here with the kind permission of the Institute.

THE AMERICAN INSTITUTE OF ARCHITECTS

AIA Document A107

The Standard Form of Agreement Between Owner and Contractor

Short Form Agreement for Small Construction Contracts

Where the Basis of Payment is a

STIPULATED SUM

For other contracts the AIA issues Standard Forms of Owner-Contractor Agreements and Standard General Conditions of the Contract for Construction for use in connection therewith.

This document has important legal consequences; consultation with an attorney is encouraged with respect to its completion or modification.

AGREEMENT

made this *first* day of *January* in the year Nineteen Hundred and *seventy-eight*

BETWEEN *Elizabeth Jones, 421 Center Street, Middleville, Illinois, hereinafter called* the Owner, and

Solaracme Contracting Co. the Contractor. *18 Main Street, Middleville, Illinois*

you can buy this form – and many others – at large stationery shops... or you can write to the AIA

The Owner and Contractor agree as set forth below.

1

ARTICLE 1
THE WORK

The Contractor shall perform all the Work required by the Contract Documents for A solar heating
(Here insert the caption descriptive of the Work as used on other Contract Documents.)
system for the house of the owner, as des-
cribed in a Request for Bids dated August
18th, 1977, and in the proposal of the contractor dated September 21, 1977.

ARTICLE 2
~~ARCHITECT~~

~~The Architect for this project is~~

ARTICLE 3
TIME OF COMMENCEMENT AND COMPLETION

The Work to be performed under this Contract shall be commenced within 7 days of the
date of this agreement
and completed within 120 calendar days thereafter.

ARTICLE 4
CONTRACT SUM

The Owner shall pay the Contractor for the performance of the Work, subject to additions and deductions by Change
Order as provided in the General Conditions, in current funds, the Contract Sum of
(State here the lump sum amount, unit prices, or both, as desired.)
Seven thousand two hundred ninety-three
dollars ($ 7293.00)

ARTICLE 5
PROGRESS PAYMENTS

Based upon Applications for Payment submitted to the ~~Architect~~ Owner by the Contractor ~~and Certificates for Payment~~ ~~issued by the Architect,~~ the Owner shall make progress payments on account of the Contract Sum to the Contractor as follows:

monthly, on or about the 5th day of each month, less 10% of each application amount, such withheld funds to be retained by the Owner until final payment.

ARTICLE 6
FINAL PAYMENT

The Owner shall make final payment *fourteen* days after completion of the Work, provided the Contract be then fully performed, subject to the provisions of Article 17 of the General Conditions.

ARTICLE 7
ENUMERATION OF CONTRACT DOCUMENTS

The Contract Documents are as noted in Paragraph 8.1 of the General Conditions and are enumerated as follows:
(List below the Agreement, Conditions of the Contract (General, Supplementary, and other Conditions), Drawings, Specifications, Addenda and accepted Alternates, showing page or sheet numbers in all cases and dates where applicable.)

(see Article 1)

AIA DOCUMENT A107 • SMALL CONSTRUCTION CONTRACT • SEPTEMBER 1966 EDITION • AIA®
© 1966 THE AMERICAN INSTITUTE OF ARCHITECTS, 1735 N.Y. AVE., N.W., WASH., D.C. 20006

3

ARTICLE 8
CONTRACT DOCUMENTS

8.1 The Contract Documents consist of this Agreement (which includes the General Conditions), Supplementary and other Conditions, the Drawings, the Specifications, all Addenda issued prior to the execution of this Agreement, all amendments, Change Orders, and written interpretations of the Contract Documents issued by the Architect. These form the Contract and what is required by any one shall be as binding as if required by all. The intention of the Contract Documents is to include all labor, materials, equipment and other items as provided in Paragraph 11.2 necessary for the proper execution and completion of the Work and the terms and conditions of payment therefor, and also to include all Work which may be reasonably inferable from the Contract Documents as being necessary to produce the intended results.

8.2 The Contract Documents shall be signed in not less than triplicate by the Owner and the Contractor. If either the Owner or the Contractor do not sign the Drawings, Specifications, or any of the other Contract Documents, the Architect shall identify them. By executing the Contract, the Contractor represents that he has visited the site and familiarized himself with the local conditions under which the Work is to be performed.

8.3 The term Work as used in the Contract Documents includes all labor necessary to produce the construction required by the Contract Documents, and all materials and equipment incorporated or to be incorporated in such construction.

ARTICLE 9
ARCHITECT

9.1 The Architect will provide general administration of the Contract and will be the Owner's representative during the construction period.

9.2 The Architect shall at all times have access to the Work wherever it is in preparation and progress.

9.3 The Architect will make periodic visits to the site to familiarize himself generally with the progress and quality of the Work and to determine in general if the Work is proceeding in accordance with the Contract Documents. On the basis of his on-site observations as an architect, he will keep the Owner informed of the progress of the Work, and will endeavor to guard the Owner against defects and deficiencies in the Work of the Contractor. The Architect will not be required to make exhaustive or continuous on-site inspections to check the quality or quantity of the Work. The Architect will not be responsible for construction means, methods, techniques, sequences or procedures, or for safety precautions and programs in connection with the Work, and he will not be responsible for the Contractor's failure to carry out the Work in accordance with the Contract Documents.

9.4 Based on such observations and the Contractor's Applications for Payment, the Architect will determine the amounts owing to the Contractor and will issue Certificates for Payment in accordance with Article 17.

9.5 The Architect will be, in the first instance, the interpreter of the requirements of the Contract Documents. He will make decisions on all claims and disputes between the Owner and the Contractor. All his decisions are subject to arbitration.

9.6 The Architect has authority to reject Work which does not conform to the Contract Documents and to stop the Work, or any portion thereof, if necessary to insure its proper execution.

ARTICLE 10
OWNER

10.1 The Owner shall furnish all surveys.

10.2 The Owner shall secure and pay for easements for permanent structures or permanent changes in existing facilities.

10.3 The Owner shall issue all instructions to the Contractor through the Architect.

ARTICLE 11
CONTRACTOR

11.1 The Contractor shall supervise and direct the Work, using his best skill and attention. The Contractor shall be solely responsible for all construction means, methods, techniques, sequences and procedures and for coordinating all portions of the Work under the Contract.

11.2 Unless otherwise specifically noted, the Contractor shall provide and pay for all labor, materials, equipment, tools, construction equipment and machinery, water, heat, utilities, transportation, and other facilities and services necessary for the proper execution and completion of the Work.

11.3 The Contractor shall at all times enforce strict discipline and good order among his employees, and shall not employ on the Work any unfit person or anyone not skilled in the task assigned to him.

11.4 The Contractor warrants to the Owner and the Architect that all materials and equipment incorporated in the Work will be new unless otherwise specified, and that all Work will be of good quality, free from faults and defects and in conformance with the Contract Documents. All work not so conforming to these standards may be considered defective.

11.5 The Contractor shall pay all sales, consumer, use and other similar taxes required by law and shall secure all permits, fees and licenses necessary for the execution of the Work.

11.6 The Contractor shall give all notices and comply with all laws, ordinances, rules, regulations, and orders of any public authority bearing on the performance of

AIA DOCUMENT A107 • SMALL CONSTRUCTION CONTRACT • SEPTEMBER 1966 EDITION • AIA®
©1966 THE AMERICAN INSTITUTE OF ARCHITECTS, 1735 N.Y. AVE., N.W., WASH., D.C. 20006

4

169

the Work, and shall notify the ~~Architect~~ [Owner] if the ~~Drawings~~ and Specifications are at variance therewith.

11.7 The Contractor shall be responsible for the acts and omissions of all his employees and all Subcontractors, their agents and employees and all other persons performing any of the Work under a contract with the Contractor.

~~11.8 The Contractor shall furnish all samples and shop drawings as directed for approval of the Architect for conformance with the design concept and with the information given in the Contract Documents. The Work shall be in accordance with approved samples and shop drawings.~~

11.9 The Contractor at all times shall keep the premises free from accumulation of waste materials or rubbish caused by his operations. At the completion of the Work he shall remove all his waste materials and rubbish from and about the Project as well as his tools, construction equipment, machinery and surplus materials, and shall clean all glass surfaces and shall leave the Work "broom clean" or its equivalent, except as otherwise specified.

11.10 The Contractor shall indemnify and hold harmless the Owner ~~and the Architect~~ and [her] agents and employees from and against all claims, damages, losses and expenses including attorneys' fees arising out of or resulting from the performance of the Work, provided that any such claim, damage, loss or expense (a) is attributable to bodily injury, sickness, disease or death, or to injury to or destruction of tangible property (other than the Work itself) including the loss of use resulting therefrom, and (b) is caused in whole or in part by any negligent act or omission of the Contractor, any Subcontractor, anyone directly or indirectly employed by any of them or anyone for whose acts any of them may be liable, regardless of whether or not it is caused in part by a party indemnified hereunder. In any and all claims against the Owner ~~or the Architect~~ or any of [her] agents or employees by any employee of the Contractor, any Subcontractor, anyone directly or indirectly employed by any of them or anyone for whose acts any of them may be liable, the indemnification obligation under this Paragraph 11.10 shall not be limited in any way by any limitation on the amount or type of damages, compensation or benefits payable by or for the Contractor or any Subcontractor under workmen's compensation acts, disability benefit acts or other employee benefit acts. ~~The obligations of the Contractor under this Paragraph 11.10 shall not extend to the liability of the Architect, his agents or employees arising out of (1) the preparation or approval of maps, drawings, opinions, reports, surveys, Change Orders, designs or specifications, or (2) the giving of or the failure to give directions or instructions by the Architect, his agents or employees provided such giving or failure to give is the primary cause of the injury or damage.~~

ARTICLE 12
SUBCONTRACTS

12.1 A Subcontractor is a person who has a direct contract with the Contractor to perform any of the Work at the site.

12.2 Prior to the award of the Contract the Contractor shall furnish to the ~~Architect~~ [Owner] in writing a list of the

names of Subcontractors proposed for the principal portions of the Work. The Contractor shall not employ any Subcontractor to whom the ~~Architect or~~ the Owner may have (reasonable) objection. The Contractor shall not be required to employ any Subcontractor to whom he has a reasonable objection. Contracts between the Contractor and the Subcontractor shall be in accordance with the terms of this Agreement and shall include the General Conditions of this Agreement insofar as applicable.

ARTICLE 13
SEPARATE CONTRACTS

The Owner has the right to let other contracts in connection with the Work and the Contractor shall properly cooperate with any such other contractors.

ARTICLE 14
ROYALTIES AND PATENTS

The Contractor shall pay all royalties and license fees. The Contractor shall defend all suits or claims for infringement or any patent rights and shall save the Owner harmless from loss on account thereof.

ARTICLE 15
ARBITRATION

All claims or disputes arising out of this Contract or the breach thereof shall be decided by arbitration in accordance with the Construction Industry Arbitration Rules of the American Arbitration Association then obtaining unless the parties mutually agree otherwise. Notice of the demand for arbitration shall be filed in writing with the other party to the Contract and with the American Arbitration Association and shall be made within a reasonable time after the dispute has arisen.

ARTICLE 16
TIME

16.1 All time limits stated in the Contract Documents are of the essence of the Contract.

16.2 If the Contractor is delayed at any time in the progress of the Work by changes ordered in the Work, by labor disputes, fire, unusual delay in transportation, unavoidable casualties, causes beyond the Contractor's control, or by any cause which ~~the Architect may determine~~ justifies the delay, then the Contract Time shall be extended ~~by Change Order for such reasonable time as the Architect may determine.~~

ARTICLE 17
PAYMENTS

17.1 Payments shall be made as provided in Article 5 of this Agreement.

17.2 Payments may be withheld on account of (1) defective Work not remedied, (2) claims filed, (3) failure of the Contractor to make payments properly to Subcontractors or for labor, materials, or equipment, (4) damage to another contractor, or (5) unsatisfactory prosecution of the Work by the Contractor.

17.3 Final payment shall not be due until the Contractor has delivered to the Owner a complete release of all liens arising out of this Contract or receipts in full

5

AIA DOCUMENT A107 • SMALL CONSTRUCTION CONTRACT • SEPTEMBER 1966 EDITION • AIA®
© 1966 THE AMERICAN INSTITUTE OF ARCHITECTS, 1735 N.Y. AVE., N.W., WASH., D.C. 20006

170

covering all labor, materials and equipment for which a lien could be filed, or a bond satisfactory to the Owner indemnifying him against any lien.

17.4 The making of final payment shall constitute a waiver of all claims by the Owner except those arising from (1) unsettled liens, (2) faulty or defective Work appearing after Substantial Completion, (3) failure of the Work to comply with the requirements of the Contract Documents, or (4) terms of any special guarantees required by the Contract Documents. The acceptance of final payment shall constitute a waiver of all claims by the Contractor except those previously made in writing and still unsettled.

ARTICLE 18
PROTECTION OF PERSONS AND PROPERTY

The Contractor shall be responsible for initiating, maintaining, and supervising all safety precautions and programs in connection with the Work. He shall take all reasonable precautions for the safety of, and shall provide all reasonable protection to prevent damage, injury or loss to (1) all employees on the Work and other persons who may be affected thereby, (2) all the Work and all materials and equipment to be incorporated therein, and (3) other property at the site or adjacent thereto. He shall comply with all applicable laws, ordinances, rules, regulations and orders of any public authority having jurisdiction for the safety of persons or property or to protect them from damage, injury or loss. All damage or loss to any property caused in whole or in part by the Contractor, any Subcontractor, any Subcontractor or anyone directly or indirectly employed by any of them, or by anyone for whose acts any of them may be liable, shall be remedied by the Contractor, except damage or loss attributable to faulty ~~Drawings or~~ Specifications or to the acts or omissions of the Owner ~~or Architect~~ or anyone employed ~~by either of~~ **him** ~~either of~~ **her** or for whose acts ~~either of~~ **she** may be liable but which are not attributable to the fault or negligence of the Contractor.

ARTICLE 19
CONTRACTOR'S LIABILITY INSURANCE

The Contractor shall purchase and maintain such insurance as will protect him from claims under workmen's compensation acts and other employee benefit acts, from claims for damages because of bodily injury, including death, and from claims for damages to property which may arise out of or result from the Contractor's operations under this Contract, whether such operations be by himself or by any Subcontractor or anyone directly or indirectly employed by any of them. This insurance shall be written for not less than any limits of liability specified as part of this Contract, or required by law, whichever is the greater, and shall include contractual liability insurance as applicable to the Contractor's obligations under Paragraph 11.10. Certificates of such insurance shall be filed with the Owner.

ARTICLE 20
OWNER'S LIABILITY INSURANCE

The Owner shall be responsible for purchasing and maintaining his own liability insurance and, at his op-

tion, may maintain such insurance as will protect him against claims which may arise from operations under the Contract.

ARTICLE 21
PROPERTY INSURANCE

21.1 Unless otherwise provided, the Owner shall purchase and maintain property insurance upon the entire Work at the site to the full insurable value thereof. This insurance shall include the interests of the Owner, the Contractor, Subcontractors and Sub-subcontractors in the Work and shall insure against the perils of Fire, Extended Coverage, Vandalism and Malicious Mischief.

21.2 Any insured loss is to be adjusted with the Owner and made payable to the Owner as trustee for the insureds, as their interests may appear, subject to the requirements of any mortgagee clause.

21.3 The Owner shall file a copy of all policies with the Contractor prior to the commencement of the Work.

21.4 The Owner and Contractor waive all rights against each other for damages caused by fire or other perils to the extent covered by insurance provided under this paragraph. The Contractor shall require similar waivers by Subcontractors and Sub-subcontractors.

ARTICLE 22
CHANGES IN THE WORK

22.1 The Owner without invalidating the Contract may order Changes in the Work consisting of additions, deletions, or modifications, the Contract Sum and the Contract Time being adjusted accordingly. All such Changes in the Work shall be authorized by written Change Order signed by the Owner. ~~or the Architect as his duly authorized agent.~~

22.2 The Contract Sum and the Contract Time may be changed only by Change Order.

22.3 The cost or credit to the Owner from a Change in the Work shall be determined by mutual agreement before executing the Work involved.

ARTICLE 23
CORRECTION OF WORK

The Contractor shall correct any Work that fails to conform to the requirements of the Contract Documents where such failure to conform appears during the progress of the Work, and shall remedy any defects due to faulty materials, equipment or workmanship which appear within a period of one year from the Date of ~~Substantial~~ Completion of the Contract or within such longer period of time as may be prescribed by law or by the terms of any applicable special guarantee required by the Contract Documents. The provisions of this Article 23 apply to Work done by Subcontractors as well as to Work done by direct employees of the Contractor.

ARTICLE 24
TERMINATION BY THE CONTRACTOR

If ~~the Architect fails to issue a Certificate of Payment for a period of thirty days through no fault of the Contractor, or if~~ the Owner fails to make payment thereon for a period of thirty days, the Contractor may, upon

AIA DOCUMENT A107 • SMALL CONSTRUCTION CONTRACT • SEPTEMBER 1966 EDITION • AIA®
© 1966 THE AMERICAN INSTITUTE OF ARCHITECTS, 1735 N.Y. AVE., N.W., WASH., D.C. 20006

6

171

seven days' written notice to the Owner and the Architect, terminate the Contract and recover from the Owner payment for all Work executed and for any proven loss sustained upon any materials, equipment, tools, and construction equipment and machinery including reasonable profit and damages.

ARTICLE 25
TERMINATION BY THE OWNER

If the Contractor defaults or neglects to carry out the Work in accordance with the Contract Documents or fails to perform any provision of the Contract, the Owner may, after seven days' written notice to the Contractor and without prejudice to any other remedy he may have, make good such deficiencies and may deduct the cost thereof from the payment then or thereafter due the Contractor or, at his option, may terminate the Contract and take possession of the site and of all materials, equipment, tools, and construction equipment and machinery thereon owned by the Contractor and may finish the Work by whatever method he may deem expedient, and if the unpaid balance of the Contract Sum exceeds the expense of finishing the Work, such excess shall be paid to the Contractor, but if such expense exceeds such unpaid balance, the Contractor shall pay the difference to the Owner.

don't leave blank spaces in contracts! draw diagonal lines.

This Agreement executed the day and year first written above.

OWNER _Elizabeth Jones_

CONTRACTOR _John Bitz, President_
Silvacme Contracting Co.

_make sure you get
a copy of the contract!_

8

173

Here's a sample contractor's agreement...

Poor identification of owner; no identification of contractor!

We propose to furnish and install a Solaracme Heating system at **421 Center Street, Middleville, Ill** in accordance with the following conditions and specifications:

EQUIPMENT

Solaracme collector - air - 420 sq. ft.
Solaracme blower
Solaracme control system
insulated rock chamber
Solaracme collector - liquid - 38 sq. ft.
insulated 80 gallon storage tank

LOCATION OF EQUIPMENT

1. Suitable space and access for this installation is to be provided by you.

DUCT WORK

1. Ductwork to be installed by us will be designated, fabricated and installed in accordance with American Society of Heating, Refrigerating and Air Conditioning Engineers standards, within the limits of existing installation conditions.

This can lead to big extra if it's discovered later that a beam is in the way

Too loose.

RESPONSIBILITY

The following responsibilities will be assumed by you or us as indicated:

Delivery, uncrating, erection	**US**	
Equipment Foundation	**US**	
Ductwork (as described)	**US**	Wiring from Panel to Conditioner - **YOU**
Duct Insulation	**US**	Cutting holes — **US**
Supply Outlets	**US**	Patching — **US**
Return Grilles	**US**	Redecorating - **YOU**
Adequate Electric Service - **YOU**		Plumbing within — **US**
		Pipe Insulation — **US**

Watch out for these!

Who are us?
Who am you?

174

Pretty weak; watch out!

PLUMBING Heat exchanger for domest. h. w. All pipe and fittings copper.

WIRING As required.

MISCELLANEOUS

We will conform to the requirements of your Request for Bids letter of August 18, 1977.

Good! This can really protect you if you wrote the request carefully.

WORKMANSHIP

1. Our work will be performed in the highest workmanlike manner and will comply with existing governing codes and regulations.

WARRANTY AND SERVICE

1. After installation a qualified representative will start, test, and provide instructions on the use of the equipment.
2. All equipment, material and labor furnished by us will bear a one year warranty from date of installation against defects in workmanship and material. In addition, the collectors, controls and heat exchanger are protected in accordance with the Solaracme Five Year Protection Plan.
3. Service under this warranty, except for emergency calls, will only be provided during normal working hours and does not include filter or fuse replacement.

Where is this little gem described?

GENERAL

1. During installation we shall take all reasonable precautions to avoid injury to persons and damage to property.
2. We shall not be liable for damages resulting from the use and/or installation of the equipment specified herein.
3. Title to the equipment will remain in us until all sums due us have been paid.
4. We shall have the right to transfer any or all notes held hereunder, and title or right of possession will pass to the legal holder.
5. Any equipment or labor in addition to that required by this proposal will be paid for by you as an extra at our normal rates.
6. It is understood that this proposal sets forth our entire agreement.
7. This proposal will become a contract between us if accepted by you and thereafter approved in writing by our duly authorized representative.

INSTALLATION SCHEDULE

1. The equipment will be ready for *installation* about **75** days from the date of our approval of this contract.

PRICE AND TERMS

NET CASH PRICE:
One half now; half at completion. ($**7293** 00)

Respectfully submitted

By *John Butz*

Date *1/1/78*

ACCEPTANCE

This proposal is accepted

By *Elizabeth Jones*

Date *January 1, 1978*

DEALER APPROVAL

This contract is approved

Dealer: *Solar City*

By *[signature]*

Title: *Manager*

Date *December 19, 1977*

What?
Who is responsible?

Bad business, this...

Do you know what these are? Is there no limit?

176

Is your head spinning after reading all that? Has your solar heating project begun to look like a mountain? If it has, take a deep breath and then skim back over that contract again. You'll see that it's really quite straightforward; no more complicated than the instructions that came with your washing machine. We're not trying to depress or confuse you; we simply want to make you aware of the kinds of terms often written into small construction contracts. In a few minutes, when you read the sample of a *contractor's* contract, you'll see how comforting the protections in the first one are.

Now, as you look at the contractor's form that follows the contractor's contract, remember that it's a lot better to think about all these things now than to think about them for the first time when something suddenly goes wrong.

It's a far cry from the AIA contract, isn't it? Now do you see why we wanted you to examine them both? We're tempted to make a list of all the points on which the contractor's contract is weaker than the AIA contract, but you, a person discriminating enough to buy this book, are capable of comparing the two documents without our help, and we urge you to do it. The very act of reviewing all those terms one more time will bring to your mind images of all the troubles others have had, troubles which led, ultimately, to the wording of the terms themselves, and you will be able to decide how much protection you want to have when you take your solar step.

A subtle change occurs after two people have signed a contract. They put down their pens, they look up at each other, and they see horns and tails appearing. Owner and contractor are now legally adversaries, a relationship that is all too often construed as war (with you-can-imagine-what-kind-of results).

We believe that the first few days of such a relationship are often the most important ones of all, for if the owner and the contractor do begin to regard each other as enemies they will begin to act that way, keeping score from the first, building lawsuits. On the other hand, if they start out by trying to help one another, *and really work at it,* even when an unkind remark might seem more appropriate, the contractor will be able to make a little more money on the job and the owner will get a little better installation for his money.

What does it cost you to help carry some of the contractor's materials to a more convenient place? What does it cost you to do a little extra cleanup work before the contractor returns, to ask him about his plans for expanding his business, or to quietly replace the pane of glass accidentally broken by his new apprentice? We've seen people turn volatile contractors into tame pussycats—and friends—in these ways. And isn't this what civilization is all about?

We think so, but in having to move on, now, to another subject, we may very well seem hypocritical, having just spouted all that stuff about goodness and love, and now turning around and talking as if you should trust no one, but that's the way it's going to be, so here we go.

STEP 4: DON'T OVERPAY HIM DURING CONSTRUCTION

Most small construction contracts are completed so quickly that one payment, at the end, is all that's needed, but some solar installations, in both existing and new buildings, are so dependent upon other contractors or upon the progress of sometimes tricky alteration work that monthly progress payments become necessary. Traditionally, in construction contracts, after the amount of a monthly progress payment has been approved, a certain amount, usually 10%, is withheld from each payment, and the withheld sums are retained by the owner until final payment. You probably noticed that provision in the AIA contract. The contractor will often balk at this (he likes to get what he considers to be his money as soon as possible), but don't be too easily swayed; he will no doubt withhold an equal amount from his subcontractors and suppliers.

The withheld money constitutes added protection for you in case the contractor fails to complete the work, and, on lengthy jobs (as the withheld funds accumulate) you can be a little more certain, because of the retainage, of seeing the early completion of all those dangling little pain-in-the-neck items that contractors dearly hate to come all the way back for. They always come back if enough money is involved.

Rare is the job on which every last bolt is nutted and every bit of leftover material removed; you must make a judgment as to how much the heartburn of a dispute over such things is worth. Often, it's simpler to take care of them yourself. (If we seem to be taking the side of the contractor in all these matters, remember that this book is written principally for homeowners. If our audience were made up primarily of solar installers we'd be telling them, in even stronger terms, to give you, the owner, a break.) In the field of human relations there's nothing like a mixture of fairness, cooperation, and quiet strength, the last of which brings us to the final step of the big five with which we started this chapter.

STEP 5: KEEP A WRITTEN RECORD OF EVERYTHING

You may at times have been successful in winning arguments by screaming and yelling, but when the chips are down it's what the record shows that counts. It's so easy to keep a little notebook by your telephone. It's so easy to record in it, while each event is fresh in your mind, the things that happen as your solar heating installation moves along.

Here are two samples from such a log, one before the contract was signed . . .

9/19/77	Solarajax rep still doesn't answer. Too late now to bid
9/21/77	We got 3 bids! One (solaracme's) is less than the $7500 we'd expected to pay. $7293.00
9/23/77	Mr. Butz came after supper to discuss the job. He told us he never fails to keep appointments and said our job would

. . . and one afterward:

1/29/78	We offered to let Mr. Butz store his material in our garage.
2/21/78	Contractors _finally_ showed up, took measurements and left, couldn't or wouldn't say when collectors will arrive.
2/28/78	Mr. Butz phoned us at 6:30 this morning to make sure I'd be here all day. Never showed up. While we waited we cleared out a big space in the garage for the

It is of course difficult, perhaps even beyond human power, to write such a log without slanting it a bit in favor of one's self. But it will be much stronger and more useful—and, of course, fairer—if bias is kept out of it.

There's no need to let the contractor know you're logging his movements, not at first, anyway. When, however, a disagreement arises, the record will help you exhibit your diplomatic skills. It is far, far better to negotiate in a spirit of cooperation and friendship than in anger. Anger usually flares highest when there's little substance to a claim. And, once sides have been taken, it's very hard to restore peaceful relations again.

Suppose you notice your contractor installing aluminum collector plates when you felt certain that copper had been agreed upon. And suppose (shame on you!) you had forgotten to put in the contract anything having to do with the basic materials to be used. But suppose that from among the notes in your log for the day you reviewed the bids you find something like this: "Mr. Butz said we shouldn't use aluminum collector plates; said he always uses copper ones and double glazing, but did say he might want to use long-life plastic glazing (double) instead of glass on our panels. We told him that was all right with us."

If, after you've mentioned your understanding of this to him, he still disagrees, you can then show him your notes of the earlier meeting, and offer to send him a copy of them. That's often enough to end such a dispute, but if in this case it does not, no arguments or shouts need be involved. The written record is what counts, whether the disagreement ends at the kitchen table over a handshake and a cup of coffee or in the small-claims court over a judicial settlement.

Without malice, simply record the good along with the bad and let the chips fall where they may. You'll be happier for having done it that way.

Now you're all set to go out there and do business with the best of them. Just picture us as standing at your sides, whispering encouragement into your ears as you enter your solar adventure. Keep your perspective. Look beyond the immediate to the long-range goals of much lower fuel bills and, beyond that, to our nation's return to energy-sanity, and to the vital role you'll be playing in that return.

P.S. by Malcolm Wells

Was my face red!

After reading the preceding chapter, the editor reminded me that I hadn't said a word about dealing with architects! Being one myself, I'd never had occasion to deal with others, to sit on the client's side of the table, so the subject had never occurred to me as I worked on this chapter.

I was so embarrassed by the editor's observation I almost offered to let Irwin handle the subject, but I knew that he would be too kind to say what needed to be said about the members of my profession, and I knew that, being an architect, I was perhaps better equipped to destroy the worst of the architectural myths.

Most people who've never dealt with us before have almost no idea what goes on in our offices or what kinds of fees we charge for our services. The timidity with which some clients first enter my office is a source of great embarrassment to me. I don't like seeming to be what their preconceptions say I am. Intentionally or unintentionally—I'm not sure which—architects have created a public facade that looks both expensive and exclusive, while in fact most of us, having been caught completely unprepared by the world energy-resources-environment crisis, are slashing fees and chasing commissions as we go through the leanest years in memory. Offices are closing, draftsmen are walking the streets, and many of us are looking for someone—the government, the Arabs, the housing industry, anyone—to come along and bail us out. In the good old days, a country club membership and a ready line of architectural patter were all we needed to keep construction projects coming into our offices. Most of us have yet to discover that it is we who are to blame for a large proportion of the mess. We don't want to face the fact that almost every building we ever designed was, and still is, both destructive and wasteful. We used too much glass, too much paving, too little insulation, energy-gulping heating and air-conditioning equipment, the worst kind of landscaping, terrible drainage, no solar energy, no wind power, no sewerless toilets, few operable windows, and no waste management. No wonder we're now in trouble!

But, as bad as the architects' record is, we do have certain residual skills. We're pretty good problem solvers—provided we know what the real problem is. Having been trained for years to think three-dimensionally in considering the dozens of factors influencing each design decision, we're pretty good at rearranging things functionally. We're flexible. Chances are that if you can't think of a single way to add a wing to one end of your house, an architect can think of about twenty-three. We've learned to push things around quickly in our minds, short-cutting our way to design solutions by building, demolishing, and then reconstructing most of our ideas before they're ever put on paper.

Our other attribute is construction experience. Materials, workmanship, expansion, contraction, time, weather, contractors, contracts—we soon get a pretty good picture of the world of building.

A young couple now building a solar house confessed to having been extremely hesitant about involving an architect in its design. The whole idea sounded to them like losing control of their dream, like putting it in the hands of someone who'd charge God-only-knew what kind of fees for services that might well run the construction costs up out of sight.

What a sad and familiar story! It's the ground upon which far too many architects and clients meet. Not the best way to launch an important relationship.

The good news, of course, is how happy the young man and woman are with their architect. At their very first meeting he told them exactly what his fees would be, step by step, throughout the job. Later on, he found many ways to simplify and thereby reduce the cost of the project; he helped them improve their design, and he was able to answer most of their questions about the uses of various materials. Then, when it was time for construction, he helped them through the bidding process, kept out the fly-by-night contractors, and got the job launched within the budget. Later, when the owners felt they could administer the project during its last few months, they paid the architect for his services up to that point, and parted on the best of terms. Now, whenever they have a construction problem or a question, they feel relaxed about consulting him. If it's a simple problem he doesn't charge for his time; if it's not so simple he charges the hourly rate worked out at the first meeting.

Hourly rates for architects range from as little as $15 to as much as $50. The nice part is that in an hour or two an architect can often suggest design changes or construction tips that will save his clients far more than the amount of his fee. If he's aware of life-cycle costs he may do even better.

Do I sound like an architect selling architectural services? I don't mean to. I'm not at all sure that an architect should be involved in the solarizing of an existing building, not unless he has great interest and experience in the subject, and the homeowner feels incapable of handling the job himself.

Remember, architects are not equipment designers. In fact, most states forbid architects to practice any kind of engineering. If you need professional help in the design of solar hardware, see a mechanical engineer. If you need design help related to the space, the appearance, or the finishes of a building, consult an architect. And, if the project is large enough to justify it, hire an architect and have him in turn engage—under his basic fee—the engineers needed to handle those basic specialities—structural, mechanical, electrical—traditionally outside the architect's own specialties.

Now, before we leave this long architectural postscript, let us look together at this standard agreement form on the next page.*

"But," you may be thinking, "all I wanted was a simple little solar system and this guy is recommending architects, engineers, and lawyers."

No, I'm not. I want you to make your project as simple and as cheap as possible. But I also want you to stay out of trouble. That's why you should know what all the options are. Then you can make informed decisions with more confidence.

*Reprinted here with the kind permission of the Institute.

this is not always necessary.

THE AMERICAN INSTITUTE OF ARCHITECTS

AIA Document B151

Abbreviated Form of Agreement Between Owner and Architect For Construction Projects of Limited Scope

THIS DOCUMENT HAS IMPORTANT LEGAL CONSEQUENCES; CONSULTATION WITH AN ATTORNEY IS ENCOURAGED WITH RESPECT TO ITS COMPLETION OR MODIFICATION.

AGREEMENT made this _eleventh_ day of _November_ in the year of Nineteen Hundred and _Seventy-seven._

BETWEEN the Owner: _Elizabeth Jones_

and the Architect: _Terrence Montague Belford III_

For the following Project:
(Include detailed Project location and scope.)

The design of a solar heating system for an existing house at 421 Center Street, Middleville, Illinois.

The Owner and the Architect agree as set forth below.

I. THE ARCHITECT shall provide professional services for the Project in accordance with the Terms and Conditions of this Agreement.

II. THE OWNER shall compensate the Architect, in accordance with the Terms and Conditions of this Agreement.

  A. FOR SERVICES, as described in Article 1, compensation shall be _Five hundred dollars._ ← _this could be a % of the construction cost, or even a straight hourly rate_

  B. AN INITIAL PAYMENT OF _None_ dollars ($ _____) shall be made upon execution of this Agreement and credited to the Owner's account.

  C. FOR REIMBURSABLE EXPENSES, amounts expended as defined in Paragraph 4.3.

  D. IF PROJECT SCOPE or Article 1 services are changed materially, or if services covered by this Agreement have not been completed within (_six_) months of the date hereof, the amount of compensation shall be subject to renegotiation.

1

184

TERMS AND CONDITIONS OF AGREEMENT BETWEEN OWNER AND ARCHITECT

ARTICLE 1

ARCHITECT'S SERVICES

The Architect's Services consist of the four phases described below and include normal structural, mechanical and electrical engineering services and any other services included in Article 11 as related to a single Stipulated Sum Construction Contract. The extent of the Architect's duties and responsibilities and the limitations of his authority as assigned hereunder shall not be modified without his written consent.

DESIGN PHASE

1.1 The Architect shall prepare Design Studies consisting of drawings and other documents for approval by the Owner, and shall submit to the Owner a Statement of Probable Construction Cost.

CONSTRUCTION DOCUMENTS PHASE

1.2 The Architect shall prepare from the approved Design Studies, Drawings and Specifications setting forth in detail the requirements for the Project, and shall submit an adjusted Statement of Probable Construction Cost.
1.2.1 The Architect shall assist the Owner in filing the required documents for the approval of governmental authorities having jurisdiction over the Project.

BIDDING OR NEGOTIATION PHASE

1.3 The Architect, following the Owner's approval of the Construction Documents and of the adjusted Statement of Probable Construction Cost, shall assist the Owner in obtaining bids and in awarding the Construction Contract.

CONSTRUCTION PHASE

1.4 The Construction Phase will commence with the award of the Construction Contract and will terminate when the final Certificate for Payment is issued to the Owner.
1.4.1 The Architect shall provide general Administration of the Construction Contract, as set forth below.
1.4.2 All of the Owner's instructions to the Contractor shall be issued through the Architect. The Architect shall prepare all Change Orders.
1.4.3 The Architect shall make periodic visits to the site to familiarize himself generally with the progress and quality of the Work and to determine in general if the Work is proceeding in accordance with the Contract Documents. On the basis of his on-site observations as an architect, he shall endeavor to guard the Owner against defects and deficiencies in the Work of the Contractor. The Architect shall not be required to make exhaustive or continuous on-site inspections to check the quality or quantity of the Work. The Architect shall not be responsible for construction means, methods, techniques, sequences or procedures, or for safety precautions and programs in connection with the Work, and he shall not be responsible for the Contractor's failure to carry out the Work in accordance with the Contract Documents.

1.4.4 Based on such observations at the site and on the Contractor's Applications for Payment, the Architect shall determine the amount owing to the Contractor and shall issue Certificates for Payment in such amounts. The issuance of a Certificate for Payment shall constitute a representation by the Architect to the Owner, based on the Architect's observations at the site as provided in Subparagraph 1.4.3 and the data comprising the Application for Payment, that the Work has progressed to the point indicated; that to the best of the Architect's knowledge, information and belief, the quality of the Work is in accordance with the Contract Documents (subject to an evaluation of the Work for conformance with the Contract Documents upon Substantial Completion, to the results of any subsequent tests required by the Contract Documents, to minor deviations from the Contract Documents correctable prior to completion, and to any specific qualifications stated in the Certificate for Payment); and that the Contractor is entitled to payment in the amount certified. By issuing a Certificate for Payment, the Architect shall not be deemed to represent that he has made any examination to ascertain how and for what purpose the Contractor has used the moneys paid on account of the Contract Sum.

1.4.5 The Architect shall be the interpreter of the requirements of the Contract Documents and the impartial judge of performance thereunder by both the Owner and Contractor, and shall make decisions on all claims of the Owner and Contractor relating thereto.

1.4.6 The Architect shall review and approve shop drawings, samples, and other submissions of the Contractor only for conformance with the design concept of the Project and for compliance with the information given in the Contract Documents.

1.4.7 The Architect shall conduct inspections to determine the Dates of Substantial Completion and final completion, and shall issue a final Certificate for Payment.

1.4.8 The Architect shall not be responsible for the acts or omissions of the Contractor, or any Subcontractors, or their agents or employees, or any other persons performing any of the Work.

ARTICLE 2

THE OWNER'S RESPONSIBILITIES

2.1 The Owner shall provide full information, including a complete program, regarding his requirements for the Project.
2.2 The Owner shall furnish full information about and affecting the site, including a certified land survey, and

AIA DOCUMENT B151 • ABBREVIATED OWNER-ARCHITECT AGREEMENT • FEBRUARY 1974 EDITION • AIA®
© 1974 • THE AMERICAN INSTITUTE OF ARCHITECTS, 1735 NEW YORK AVE., N.W., WASHINGTON, D.C. 20006

2

when deemed necessary by the Architect, soil test reports or the services of a soil engineer.

2.3 The Owner shall furnish laboratory tests, inspections, and reports as required by law or the Contract Documents.

2.4 The Owner shall furnish such legal, accounting and insurance counseling services necessary for the Project, and such auditing services as he may require to ascertain how the Contractor has used the money paid to him.

2.5 The information, surveys, and reports required by Paragraphs 2.1 through 2.4 inclusive shall be furnished at the Owner's expense, and the Architect shall be entitled to rely upon the accuracy and completeness thereof.

2.6 If the Owner becomes aware of any fault or defect in the Project or nonconformance with the Contract Documents, he shall give prompt written notice to the Architect.

2.7 The Owner shall furnish information required of him as expeditiously as necessary for the orderly progress of the Work.

Read! →

ARTICLE 3
CONSTRUCTION COST

3.1 The Construction Cost shall be the total cost or estimated cost to the Owner of all Work designed or specified by the Architect, which shall be determined as follows, with precedence in the order listed:

3.1.1 For completed construction, the cost of all such Work, including the cost of labor, materials and equipment furnished by the Owner and the cost of managing construction; or

3.1.2 For Work not constructed, (1) the lowest bona fide bid received from a qualified bidder for any or all of such Work, or (2) if the Work is not bid, the bona fide negotiated proposal submitted for any or all of such Work; or

3.1.3 For Work or portions of the Work for which no such bid or proposal is received, the latest Statement of Probable Construction Cost.

3.2 Construction Cost does not include the compensation of the Architect and his consultants, the cost of the land, right-of-way, or other costs which are the responsibility of the Owner in Paragraphs 2.1 through 2.4 inclusive.

3.3 The Architect cannot and does not guarantee that bids will not vary from Statements of Probable Construction Cost or other cost estimates prepared by him.

3.4 When a fixed limit of Construction Cost is established as a condition of this Agreement, it shall be in writing signed by the parties and shall include a bidding contingency of ten percent, and if it is exceeded by the lowest bona fide bid or negotiated proposal, the Owner shall (1) give written approval of an increase in such fixed limit, (2) authorize rebidding the Project within a reasonable time, or (3) cooperate in revising the Project to reduce the Probable Construction Cost. In the case of (3) the Architect, without additional charge, shall discharge his responsibility by modifying the Drawings and

Specifications, and having done so, shall be entitled to compensation in accordance with this Agreement.

ARTICLE 4
PAYMENTS TO THE ARCHITECT

4.1 An initial payment as set forth in Paragraph II is the minimum payment under this Agreement.

4.2 Payments for Services shall be made monthly, in proportion to services performed. If compensation is on the basis of a fixed fee or percentage of construction cost it shall, at the completion of each Phase, equal the following percentages of the total Compensation:

Design Phase
Construction Documents Phase   35%
Bidding or Negotiation Phase   75%
Construction Phase   80%
   100%

4.3 Payment for Reimbursable Expenses shall be made monthly. Reimbursable Expenses are in addition to compensation and include actual expenditures made by the Architect for the Project for: travel and subsistence; long distance calls; fees paid to governmental authorities; renderings and models required by the Owner; Owner authorized overtime; reproductions, postage and handling of Drawings and Specifications, excluding duplicate sets at the completion of each Phase for the Owner's review and approval.

4.4 No deductions shall be made from the Architect's compensation on account of sums withheld from payments to contractors.

4.5 If the Project is suspended for more than three months or abandoned in whole or in part, the Architect shall be paid for services performed prior to receipt of such notice from the Owner together with all termination expenses. If the Project is resumed after being suspended for more than three months, the Architect's compensation shall be subject to renegotiation.

4.6 Payments due the Architect under this Agreement shall bear interest at the legal rate commencing sixty days after date of billing.

ARTICLE 5
TERMINATION OF AGREEMENT

5.1 This Agreement may be terminated by either party upon seven days written notice should the other party fail substantially to perform in accordance with its terms through no fault of the party initiating the termination. In the event of termination due to the fault of parties other than the Architect, the Architect shall be paid his compensation for services performed to termination date, including Reimbursable Expenses, plus termination expenses.

5.2 Termination expenses are defined as Reimbursable Expenses directly attributable to termination, plus an amount computed as a percentage of the total compensation earned to the time of termination as follows:

AIA DOCUMENT B151 • ABBREVIATED OWNER-ARCHITECT AGREEMENT • FEBRUARY 1974 EDITION • AIA®
© 1974 • THE AMERICAN INSTITUTE OF ARCHITECTS, 1735 NEW YORK AVE., N.W., WASHINGTON, D.C. 20006

3

this is a surprise to many owners.

20 percent if termination occurs during the Design Phase; or

10 percent if termination occurs during the Construction Documents Phase; or

5 percent if termination occurs during any subsequent phase.

ARTICLE 6

OWNERSHIP OF DOCUMENTS

Drawings and Specification as instruments of service are and shall remain the property of the Architect whether the Project for which they are made is executed or not. They are not to be used by the Owner on other projects or extensions to this Project except by agreement in writing and with appropriate compensation to the Architect.

ARTICLE 7

SUCCESSORS AND ASSIGNS

The Owner and the Architect each binds himself, his partners, successors, assigns and legal representatives to the other party to this Agreement and to the partners, successors, assigns and legal representatives of such other party with respect to all covenants of this Agreement. Neither the Owner nor the Architect shall assign his interest in this Agreement without the written consent of the other.

ARTICLE 8

ARBITRATION

All claims, disputes and other matters in question between the parties to this Agreement, arising out of, or relating to this Agreement or the breach thereof, shall be decided by arbitration in accordance with the Construction Industry Arbitration Rules of the American Arbitration Association then obtaining unless the parties mutually agree otherwise. No arbitration, arising out of, or relating to this Agreement shall include, by consolidation, joinder or in any other manner, any additional party not a party to this Agreement except by written consent containing a specific reference to this Agreement and signed

by all the parties hereto. Any consent to arbitration involving an additional party or parties shall not constitute consent to arbitration of any dispute not described therein or with any party not named or described therein. This Agreement to arbitrate and any agreement to arbitrate with an additional party or parties duly consented to by the parties hereto shall be specifically enforceable under the prevailing arbitration law. In no event shall the demand for arbitration be made after the date when such dispute would be barred by the applicable statute of limitations. The award rendered by the arbitrators shall be final.

ARTICLE 9

EXTENT OF AGREEMENT

This Agreement represents the entire and integrated agreement between the Owner and the Architect and supersedes all prior negotiations, representations or agreements. This Agreement may be amended only by written instrument signed by both Owner and Architect.

ARTICLE 10

GOVERNING LAW

This Agreement shall be governed by the law of the principal place of business of the Architect.

ARTICLE 11

OTHER CONDITIONS OR SERVICES

Termination expenses (see 5.2) shall be limited to expenses attributable to termination and shall not include the percentages listed in Article 5.

This Agreement executed the day and year first written above.

OWNER *Elizabeth Jones*

ARCHITECT *Pierre Montague Telford III*

4

AIA DOCUMENT B151 • ABBREVIATED OWNER-ARCHITECT AGREEMENT • FEBRUARY 1974 EDITION • AIA®
© 1974 • THE AMERICAN INSTITUTE OF ARCHITECTS, 1735 NEW YORK AVE., N.W., WASHINGTON, D.C. 20006

APPENDIX/CHAPTER 7

BIBLIOGRAPHY

Be Sure Before You Sign. A pamphlet produced by the Office of Consumer Affairs.

The Blue Book Contractors Register. Published and copyrighted 1975 by Contractors Register, Inc., Elmsford, NY 10523.

Business Law. R. A. Anderson and W. A. Kumpf. 5th ed. Cincinnati: South-Western Publishing Co.

"Contracting a Home Contractor." *The Sunday Bulletin.* Common Cents, (Philadelphia, Pa.) 6 March 1977, p. 5.

"Finding and Working with a Contractor." Michael de Courcy Hinds. *House and Garden,* May 1977, pp. 110, 113, 116, 119.

Low-Cost, Energy-Efficient Shelter for the Owner and Builder. Edited by Eugene Eccli. Emmaus, Pa.: Rodale Press, 1976.

Sunlight and the Law

We were looking into our crystal ball trying to decide how to begin this chapter on solar legal considerations, when a shadow came into view . . . *"Shadows."**

> After dawn the helicopters took off from the base and headed down the valley. Their gigantic shrouds billowed as they were towed through the air.
>
> The ominous shadows passed over the streets and rooftops of the town. No one looked up even though the noise of the helicopters was deafening. On the outskirts of town the helicopters wheeled and took up stationary positions—you could see the crewmen struggling with lines as the shrouds were lowered.
>
> It was a method of nonviolent control for dissidents who were disconnecting from the power system and going solar. No solar energy if you are in the shade. The used helicopters of previous wars were now a familiar sight.
>
> Of course, some enraged people in the shadows had at first shot at the helicopters, but the helicopters were so heavily armed it was no battle.
>
> A few weeks in the shade and the solar houses were out of commission, gripped by a chill like that which you find in a deep-shaded alley, the vegetables in their greenhouses turning yellow, like grass under a board.
>
> No one could take it for long—the chilliness and the continual throbbing roar of the helicopters; the bored crewmen showering you with

*"Shadows," reprinted from *Sunspots*, a paperback book of intriguing thoughts and ideas about the sun, with the permission of the author, Steve Baer.

There's no solar energy if you're in the shade.

wrappers, orange peels, and soft-drink cans; the terrifying sight of the
machine guns.
With the rationing of fuel oil and gasoline, this extravagant use of fuel
was insult added to injury.
The local governments used the shrouds skillfully—shading a house,
a demonstration, or even, as in the case of X, a family picnic; as little as
ten minutes would often bring results.

Thought provoking and chilling, although not a likely probability here
in the good old United States. And yet—a possibility. Here more than in
most countries, there exists a world of laws, government, and politics
which often (sometimes unintentionally) restricts your ability to use
techniques such as solar energy conversion.

Before launching yourself into a project which may cost several
thousand dollars, it's wise to examine the legal restrictions which might
affect your solar installation. In this chapter we will review and catego-
rize these possible legal restrictions. Being aware of them will help you
avoid those "sudden surprises" that often haunt even the most well-
meaning projects. Basically, there are four areas of possible legal impediments:

Access to sunlight
Building Codes

Zoning laws

Other potential legal restrictions:
  tax laws
  utilities—public and private
  labor laws
  insurance laws

ACCESS TO SUNLIGHT

The frightening example with which we opened this chapter highlights the importance of free access to sunlight. But *do* we have free access, and have we always? The answer to both questions is *no*! To give you a better perspective on this subject we must look back into the pages of history.

Prior to the 1500s, a Latin phrase, *Cujus est solum, ejus est usque ad coleum et ad infernos,* had long been incorporated into the statutes of French, German, Roman, and Jewish law. Its meaning: "He who owns the soil also owns to the heavens and to the depths." The property owner was said to own all the way to the heavens, and to the very core of the earth (wherever that may have been when the earth was flat!).

Then, in 1586, a famous law case in England occurred between two parties, Bury and Pope. Mr. Bury, it seems, had built a house close to Mr. Pope's property line, and received sunlight through one window on that side of his house. Mr. Pope later decided to build a house close to the same property line; a structure that would prevent sunlight from entering Mr. Bury's window. Bury then took Pope to court claiming that Pope should not be permitted to build so close that the light at Bury's window would be cut off. The judge decided that Bury had been foolish to build his house so close to the property line, and cited the Latin maxim, which in that way became part of the English law.

Later, the maxim was modified when the Doctrine of Ancient Lights was formulated by the English courts. It held that the owner of property was entitled to receive light across his neighbor's land up to the amount he needed for reasonable use and enjoyment of his own land, if he had received that light for a very long time. This was rather vague, so a time period of twenty years was established by the Limitation Act of 1623. Later, it was modified by the English courts to twenty-seven years in the Right to Light Act of 1959.

An interesting sidenote relates to the ingenious way in which the English courts chose to define ". . . the amount [of light] he needs for

The Bury/Pope classic.

reasonable use . . .": they established the "grumble line." The grumble line was to be the position in a room at which an ordinary person reading ordinary print would grumble and turn on the artificial light! If at least half the room was between the grumble line and the window, the courts held that the light entering the room was a reasonable amount.

Because the American legal system was founded upon the English common law, many colonial courts did accept the Doctrine of Ancient Lights, but later, courts in New England repudiated it as being inconsistent with a growing and dynamic country.

So much for history; where do we stand today? A case filed in Miami in 1959 is currently cited whenever the question of the right to sunlight is raised. In the early 1950s, the Fontainbleau and Eden Roc hotels were built adjacent to one another. Later, the owners of the Fontainbleau decided to add fourteen stories at the northern end of their property. The addition would have put the Eden Roc's pool area in the

The Eden Roc/Fontainbleau case.

shade on most winter afternoons. The owners of the Eden Roc sued, claiming they had a right to continue receiving sunlight across the Fontainbleau property. The lower court agreed, and enjoined the Fontainbleau's owners from proceeding with their plans. But an appeals court said otherwise; the judge held that the Eden Roc had no rights that were being violated (because the Doctrine of Ancient Lights *is not law* in the United States) and that no cause for action existed, "regardless of the fact that the structure may have been erected partly for spite." So now you know.

Current law in the United States has established the fact that the owner of a property has a right to receive light from the area of the sky directly above that property, but has no right to receive it from across his neighbor's land. It appears, then, that at this moment of our solar infancy, your neighbor can construct a building, plant trees, put up fences, or do anything he chooses, even though it may block out your sunlight. Of course, there are many other constraints to such actions, constraints such as zoning laws, which will be discussed later in this chapter.

What legal precautions should you take before committing yourself to the installation of a solar heating system? First, familiarize yourself with your local community's zoning regulations. Check with others in your state who have done solar work. Did they run into problems? Keep the whole thing in perspective. If your property is a large one, build on the northern portion, then you'll have no problems. Just remember not to block someone else's sunlight. If you live in a community of low buildings, keep an eye on the master plan and height limitations of the town's building code, then you'll be less likely to have a high-rise surprise. On the other hand, if you live in a commercially or industrially zoned area, proceed with caution.

One final thought on the subject of access to sunlight: other legal issues may influence your right to it. The laws concerning pollution and weather modification can directly affect your rights. Long periods of rain which result from weather modification technology can directly affect your solar heating system, for obvious reasons. Less obvious is the decreased efficiency you can observe as the result of high levels of particulate matter in the air.

During a seminar and workshop* on the subject of solar energy and the law, sponsored by the American Bar Foundation and the National Science Foundation, Mr. William A. Thomas, principal investigator for the Bar Foundation, concluded his comments on this subject with the following statement:

> . . . several centuries ago the chief reason for securing right to light across a neighbor's land was interior lighting. With the advent of inexpensive electricity our concern shifted from interior lighting to the preservation of scenic vistas and other esthetic considerations. We now are refocusing our concern on light for interior lighting, heating, and other utilitarian purposes. I know there's a fine moral there somewhere.

BUILDING CODES

Building codes regulate the construction of buildings, the materials used and how they're used, where they can or can't be used, and practically every other conceivable aspect of construction, for more than 90% of the national population. In theory this is good, since the codes

*A portion of the material used in writing this chapter was taken from the official proceedings of this seminar, which was held in Arlington, Virginia, on February 10, 1975.

are designed to protect the public health and welfare. Most communities, both suburban and urban, use at least one set of building codes, or standards, to control new construction or changes to existing buildings within their jurisdictions. Most exceptions to the existence of building codes occur in small rural communities.

You may wonder why, if these codes are designed to protect the public health and welfare, we have listed them as a possible legal impediment to your solar installation. Well, the very fact that there *are* so many nonstandard "standards," discourages technological innovation. An example which occurred in a dynamically growing town in New Jersey may help you better understand the problem.

A man and his wife were delighted with the fourteen-year-old split-level they'd purchased—except for one thing; there was no fireplace. The home they had just left, although old, and heavy on maintenance, had had a log-burning fireplace in the living room. They'd used the fireplace regularly. So they decided to install one in their new home. After much thought and analysis, they decided on a heavy ceramic firepot, to be placed on a small, raised brick hearth in the corner of their living room. Their reasons were:

• The new living room had a cathedral ceiling which would make the room difficult to heat in the winter, particularly because of a floor-to-ceiling picture window on the northern wall of the room. They felt that a free-standing firepot would prove more efficient than a built-in one in warming the room.
• During their investigation they learned that the heavy-walled ceramic firepot would burn its fuel slowly and completely, and would

A heavy ceramic firepot.

retain the heat in its walls, providing warmth to the room long after
the fire had been out.

• They also learned that the ceramic pot was safer than any of the
more common sheet steel free-standing fireplaces. When in use, the
steel units get hot enough to burn a person touching them severely,
whereas the ceramic unit gets no hotter than an old-fashioned hot-
water radiator.

In spite of all they learned, their building permit (needed before the
firepot could be installed) was denied. The building inspector for the
township, which used the BOCA (Building Officials and Code
*A*dministrators) building code, could find no reference to ceramic
firepots in the code. He did, however, find free-standing steel fireplaces listed, and said *they* could be approved (even though the steel
units were more dangerous) but that the ceramic firepots would need
"certification."

This started a long and arduous investigation by the two fireplace
lovers, one which proved quite costly in terms of time and dollars, but
eventually certification was obtained. They also had to make several
submissions of drawings and sketches showing in great detail the
construction techniques to be employed. And finally, they counted
eight visits to their home by the building inspectors during the installation period.

There are several lessons which can be learned from this example,
but first we'll finish the story. The firepot has been doing its job safely
for the past six years. The owners find that it reduces their fuel bills by
about $30/month (at 1976 fuel costs) when it's used, and it does a superb
job of warming the cathedral-ceiling-type living room. The lessons
are—

1. New or innovative materials for use in building construction (or modification) are rarely found listed in the most commonly used building codes. This is a built-in obstacle to solar
installations.

2. The couple in the example felt that the building inspection
department was prejudiced against an individual (as opposed
to a professional contractor) doing this type of building modification. Their reason for this feeling was that they had learned
of the ceramic firepot from a neighbor who had had the same
unit installed by a professional contractor. Particularly

disturbing was the fact that the building permit had been quickly issued to the neighbor's contractor on the basis of only a freehand outline sketch of the installation. You'll recall that our couple needed detailed drawings, firepot certification, and many visits by the inspector, while none of this was needed by the contractor—for the same kind of job.

3. The next lesson, then, is that you must be prepared to encounter many more legal obstacles if you plan to do the job yourself, rather than have it done by professional contractors. (It has been suggested to the authors that in some communities "arrangements" exist between the building inspection department personnel and "favored" contractors. All others appear to be harrassed by "letter-of-the-code-book" inspections every inch of the way.)

4. It was also suggested to the couple by various friends that when, in a minor home-alteration job, continuous and frequent visits of a building inspector occur, it may indicate that "an expression of thanks for his concern about public safety" in the form of a gratuity, can put a quick stop to his visits.

We are not suggesting that you pay off an inspector to get approval to proceed with your solar installation. On the contrary, in fact, payoffs can only worsen bad government. We *are* suggesting, however, that the obstacles caused by building inspectors to proposed solar installations can actually go well beyond the codes themselves.

To balance the negative impressions we've given you, here's the good news. On January 1, 1977, the state of New Jersey adopted the Uniform Construction Code of New Jersey, which is similar to the BOCA code mentioned earlier. We understand that other states have already approved—or are planning—similar moves. This is good news because it replaces the multiplicity of codes which had existed in every city and township with a dependable, standardized one. Such laws are certain to make the homeowner's job simpler when dealing with his local building inspection department. They also benefit the homeowner by providing a central point of appeal for cases in which new materials or building techniques are involved. Only one authority need rule that solar components should be included in the code. Also, many of the political abuses, common on the local level, will be curtailed because of the standardization imposed by the new state law.

Another bright spot on the building code front is the effort being generated by professional groups such as the International Association of Plumbing and Mechanical Officials (IAPMO). Their Uniform Solar Energy Code, published in September 1976, provides a reasonable starting point for states or municipalities which are in search of building code information relating to solar energy work. We were particularly impressed by the attitude expressed in the foreword to their code, which, in effect, says, "We're offering a starting point—but we, too, must learn and accept changes—and we will—with your help."

Foreword to Uniform Solar Energy Code by IAPMO

The advantages of a uniform solar energy code, acceptable in the various jurisdictions, have long been recognized. The increasing needs for such a code induced the International Association of Plumbing and Mechanical Officials to pass a resolution, at their 46th Annual Business Conference, which directed the President to form a Committee to develop a basic solar energy document.

After months of concerted endeavor, this Committee, composed of representatives from Industry and Public Utility Companies, Inspectors, Plumbers, and Engineers, successfully completed the first edition of the "Uniform Solar Energy Code" which was officially adopted by the International Association of Plumbing and Mechanical Officials in September 1976.

In presenting the 1976 edition, IAPMO recognizes that the ultimate has not yet been attained. Users of this Code are respectfully urged to present whatever amendments their experience may dictate to the Association's headquarters, not later than March 1 of each year, so that by formal adoption of such, uniformity may be maintained and all will benefit. Amendments adopted by the membership will be published in revised form every third year and will keep this basic document abreast of technological development.

The use of this document is intended to provide a safe and functional solar energy system with minimum regulations. Users of the "Uniform Solar Energy Code" are urged to strive for not just the minimum good solar energy system but to keep the ultimate consumer in mind and go a step beyond and exceed the minimum. The consumer is entitled to more than a safe and sanitary solar energy system. With the exception of "high use and wear" portions of the system, the solar energy system should have the same life as other components of the building.

The "Uniform Solar Energy Code" is dedicated to all those who have unselfishly devoted their time and effort to create this Code.

If this same attitude exists on the part of other professional and government agencies (such as ASHRAE [the American Society of Heating, Refrigerating, and Air-Conditioning Engineers] and NBS [the National Bureau of Standards]) which have been hard at work preparing solar building code standards, we have a sunny future ahead. On the other hand, if codes and standards are jammed down our throats prior to the learning process—which will take several years—"the law" may once again become a major obstacle to the use of solar energy.

ZONING LAWS

Zoning is nothing more or less than *local control over land use*. Most studies indicate that local governments make more than 90% of the decisions relating to the following aspects of land use:

1. control of population density
2. location of structures on a building lot
3. height of structures
4. bulk of structures
5. use of structures
6. esthetic characteristics of structures (i.e., styles, placement, glare, materials)

Since what may be acceptable in one part of a community may be unacceptable in another, zoning can apply different rules to different areas. The impact of zoning ordinances on solar usage, therefore, is (or can be) substantial.

One of the touchiest aspects of zoning law relates to the *variance* process. A variance permits exceptions to be made to zoning laws based upon unique circumstances. Obviously, political influence can play a major role in obtaining a variance to a zoning impediment, but unfortunately most citizens wield little political clout. The solution to this problem was well stated by Richard Robbins, a professional land-use planner and deputy director and counsel of the Lake Michigan Federation.*

> There will have to be a real effort to educate the local planning boards and officials who administer these controls so they may understand the variances that are needed for solar devices. The most important necessary changes are in height, bulk, and building location regulations and in other controls to make the fitting and retrofitting of solar devices feasible under existing zoning ordinances.

A recent example will help put the zoning law impact on solar installations into perspective:

Local Florida Planners Kill Solar Collector Ban†

> Four hours after a blackout in Coral Gables, Florida, left about 3 million people without power, the local planning commission killed a proposal to ban solar collectors from the roofs of local homes.
>
> The ordinance, voted down 5 to 1, would have prohibited solar energy panels that could be seen from the street. The city's Board of Architects had claimed that the devices would damage the city's self-proclaimed "City Beautiful" image.
>
> Nine people spoke during the commissioner's hearing. Louis Spector said, "I resent anybody enacting a law taking away my God-given right to collect the power of the sun."
>
> Commissioner Al Jacobson, who later voted to reject the ordinance said, "The objection we have is to *these things* [emphasis supplied by the authors] in a city built with a Mediterranean style. This is great for a new area . . . but we don't want to let *these fellows* [he means us] go crazy. What you can do in the Rockies or the desert, you can't do here."
>
> The commission did impose a limitation that collectors seen from the street will have to be screened.
>
> Clifford Lincoln of the Florida Industry Association had written President Carter to oppose the original proposal. Lincoln told Carter that the proposal "is in direct violation of your goals for a national

---

*Excerpted from Proceedings of the Workshop on Solar Energy and the Law, February 10, 1975, Arlington, Virginia.

†This article reprinted from *Solar Utilization News.*

energy program. It's arbitrary, discriminatory, and not in the national interest."

So, as you can see, we face a tough, uphill battle with some community zoning laws before we can comfortably install solar heating systems everywhere. Although the struggle may try your patience to its limit, the battle *can* be won, as was shown in the previous news story.

The moral of this segment of the chapter is—become familiar with your community zoning laws and the people who administer them. Try to work with the officials and educate them. Sell them on the value of solar energy—and we'll all win.

OTHER POTENTIAL RESTRICTIONS

Tax Laws

Two tax mechanisms have the greatest impact on a homeowner who is planning to install solar heating. First, he must pay sales taxes, or "use" taxes, on the material used in the solar components, and then pay increased property taxes on the added value of his solar installation. In addition, tax laws can affect local businesses in several less obvious ways, such as by increasing their income tax liability because of the lower operating expense associated with their heating needs.

A government survey,* published in April 1976, reviewed state legislation relating to solar energy. It provided details and copies of all legislation proposed or passed by:

- states with acts providing property tax incentives for use of solar energy in buildings,
- states with acts providing for income or sales tax incentives for use of solar energy and other solar considerations, and
- states with acts providing support of research, development, promotion, or investigation of solar energy.

A summary of this report which names the states involved and the incentives they have proposed (or have already passed into law), is in-

*Publication NBSIR 76-1082 (NTIS PB 258-235) prepared for ERDA and HUD by Robert M. Eisenhard of NBS (National Bureau of Standards).

Utilities tap a fuel source called conservation

For 40 years the Bonneville Power Administration has promoted greater use of electricity in the Pacific Northwest as required by its congressional charter. Now, in an abrupt about-face, this federal power-marketing agency is pushing conservation instead. It has told the utilities it serves that they could save $800 million in power plant investments—half the cost of a new nuclear power plant—if they would spend just $300 million insulating their customers' electrically heated homes. Says BPA Administrator Donald P. Hodel: "The cheapest, cleanest, and quickest source for balancing our power requirements is energy conservation."

Few utility executives would go so far as Hodel, actually proposing that power companies pick up the tab for their customers' insulation, but more and more these days both gas and electricity companies are stepping up efforts to

They employ a variety of ways to help customers insulate their homes

encourage better insulation of their customers' homes. For some time the companies have been promoting insulation and other conservation steps in their advertising. Now they are beginning to help finance and install insulation, too. Some have even formed subsidiaries to sell insulation and other energy-saving products.

The benefits. Their motivation is clear enough. For electric utilities, reducing the growth in the demand for power relieves the pressure to build costly new generating stations. For gas utilities, increased insulation means less reliance on high-cost substitutes, such as liquefied natural gas or gas made from oil. Recognizing these advantages, President Carter has asked Congress to approve a proposal that would require all utilities to offer both installation and financing for a potpourri of conservation measures, including ceiling, wall, and floor insulation.

Not unexpectedly, some industry officials are opposed to utilities' getting so involved in conservation. "We do not think the utilities belong in the home-improvement and financing business," says Dudley J. Taw, president of East Ohio Gas Co. "If we were to attempt to finance all our customers' winterization needs, the bill could be $500 million—more than our total plant investment today."

But a surprising number of gas and

electric companies are beginning to disagree. In April, for example, Chicago's Commonwealth Edison Co. received approval from its state regulatory commission to lend up to $2 million to customers for insulation work. Commonwealth points out that 60% of its 2.5 million customers have air conditioning and that the insulation made possible with these loans will hold down demand during the critical summer months. "But the real dollar saving is in the long run—in the generating units we don't have to build," explains John G. Eilering, the utility's vice-president for marketing.

MONEY & ENERGY SAVERS
From Washington Natural Gas

Deep in conservation: A Seattle gas utility last year sold $3.8 million worth of materials for home insulation.

Duke Power Co. in Charlotte, N. C., concurs. Says Donald H. Denton Jr., its vice-president for marketing: "Most definitely, the dollars expended for insulation would be less than the dollars needed for generation," though he argues that lending institutions, not utilities, should put up the money. Northern Illinois Gas Co. also believes that conservation is these days a cheap "source" of energy. "The gas we'll free up through conservation will be sold to other customers, says Alan R. Johnsen, senior vice-president for operations. "It's a lot cheaper to get gas this way than through more exotic means." Other utilities with insulation financing plans include Public Service Co. of Colorado, Michigan Con-

solidated Gas Co., and the Tennessee Valley Authority.

Washington Natural Gas Co. in Seattle has gone one step further—it has moved into the insulation business. During a gas shortage in 1973, state regulators sharply cut back Washington Natural's advertising budget and ordered it to stick to a conservation theme in its remaining promotional effort. Then the company decided that it might as well capitalize on the situation.

Selling insulation. Washington Natural began retailing attic insulation and eventually introduced a full line of products for both contractors and do-it-yourselfers. Sales of these conservation products brought in $3.8 million last year and ran 30% ahead of the 1976 pace during the first quarter of this year. The company's program remains active even though gas supplies from Canada have recently increased, easing the utility's short-term supply problem. "Conservation is like sitting atop a gas well," explains its vice-president for marketing, Donald C. Navarre, who estimates that, because of it, the company has freed enough gas to serve an additional 60,000 or more customers.

Because gas is in such short supply, gas utilities have moved into insulation and other forms of conservation more aggressively than most electric utilities have. Indeed, the American Gas Assn. says 30 of its members are involved in merchandising conservation techniques in some form, and it expects the number nearly to double by yearend. One utility, Bay.State Gas Co., in Massachusetts, has even launched an experimental program to train employees to install insulation it sells.

The Rosenberg plan. But the most radical idea—the one that BPA's Hodel endorses—originated last year with William G. Rosenberg, an official with the Federal Energy Administration. He proposed that the nation's gas utilities themselves pay for insulating their customers' homes, recovering these costs by including them in their rate bases. With this scheme, utilities would view investments in conservation exactly as they view investments in new supplies and would presumably opt for whichever is cheaper.

Though the proposal now seems dead in Washington, D.C., it is beginning to get attention elsewhere around the nation. The Pennsylvania Public Service Commission may impose the Rosenberg plan on gas utilities in that state. And in Seattle, Cascade Natural Gas Corp. will soon submit to the Washington Utilities & Transportation Commission a plan very much along the lines of the Rosenberg plan.

Thus, one way or another, it seems inevitable that utilities will be taking on a greater and greater conservation role. They will either opt for it voluntarily or have it forced on them by state regulatory officials, regional power agencies, or the federal government. "Much additional generation will still be required," sums up BPA's Hodel, "but it will be less than what would have been the case without a conservation program. The energy saved by many conservation measures can cost considerably less and be achieved more quickly than [the energy supplied] by the construction of new generation." ∎

Reprinted from the July 18, 1977 issue of *Business Week* by special permission. ⓒ 1977 by McGraw-Hill, Inc.

cluded in the Appendix of this chapter. The list you'll see there is impressive, but the fact remains that in most communities, the installation of solar hardware will increase the assessed value of your property. This increase will cost you more in property tax dollars, and will therefore diminish the economic benefit which the use of solar energy can provide. If your state is not one of those which has new or proposed legislation to remove or reduce this impediment to solar usage, we strongly advise you to write to your state representatives. Ask them why your state is dragging its feet in that way.

Utilities

Laws which govern public and private utility companies are incredibly complicated, in addition to varying from one state to the next. There is no question, though, that these laws and regulations can either help or hurt the development of solar heating. As an example, the article starting on the facing page shows what *Business Week* had to say on the subject.

We have quite a way to go, wouldn't you say? Now look at the article on the page that follows. Just imagine what acid fallout might do to your rooftop solar collectors!

The major impediment to solar heating which could be caused by utility companies and state laws is nicely summed up in a statement in the April 1976 issue of the *Progressive*.* In the article, Mark Northcross examines the idea of public utilities entering the field of solar energy, and supplying equipment, installing it and maintaining it. He says:

Energy conservation (the use of solar energy) has grave implications
for the long-term interests of utilities. It means decreasing demand,
and consequently a smaller market. Solar heating and cooling offer an

*"Who Will Own the Sun?" an article by Mark Northcross, a former newspaper reporter, presently an energy planner and environmental consultant in California.

The acid fallout that rains on California

Michael J. Hart, a resident of Redondo Beach, Calif., is fuming. The white picket fence that he repainted last year is gradually turning a splotchy orange, and Hart is certain that acidic fallout from a nearby Southern California Edison Co. power plant is to blame. Hart recently sent a sawed-off section of the stained fence to the California Air Resources Board (CARB) to dramatize his complaint. "If pollution damages things like this on the outside," asks the 26-year-old airline mechanic, "what's it doing to my three kids on the inside?"

Power plants now burn oil instead of gas, and many householders are dismayed

Hart is just one of a host of Californians who have recently discovered what their East Coast brethren have known for years: Acid fallout from oil-burning power plants can stain paint, corrode metal, discolor foliage, and deteriorate boat canvas and other fabrics. "The problem is widespread from Maine to Florida and the Gulf Coast," says James B. Homolya, the Environmental Protection Agency's chief of gaseous emissions research. "It apparently occurs wherever oil is burned in large industrial boilers." Until recently most California power plants burned natural gas, but now they have switched to oil.

'Acid smut.' Sulfur-containing coal does not seem to cause the same problem. Researchers at the industry-funded Electric Power Research Institute (EPRI) in Palo Alto, Calif., which this year will spend more than $6 million on sulfate research, say that oil has metallic components such as vanadium that may serve as catalysts to form acid. Coal is more mineralized and apparently neutralizes a certain amount of acid.

Utility spokesmen themselves concede that fallout, or "acid smut," from oil is a problem. An oil-burning utility boiler is a "modest-scale sulfuric acid generator," says A. J. O'Neal, chief chemist for Long Island Lighting Co. (Lilco), of Mineola, N. Y. Apparently, 3% to 6% of the sulfur dioxide produced in boilers is oxidized to form sulfur trioxide, which adheres to ash particulate in flue gases. When these hot gases shoot up the stack into the cooler atmosphere, the sulfur trioxide reacts with airborne moisture to form sulfuric acid on the dust-like ash.

The large particles—those larger than 50 microns, or about half the diameter of a human hair—normally fall within a 3-mi. radius of the source and can cause

Californians who live near oil-burning power plants are finding blotches of 'acid smut' on all outdoors.

extensive property damage. The smaller particles, along with some free sulfuric acid, form acid mist, which can travel long distances and may fall to earth as the much-discussed "acid rain."

So far, the mist has caused the most worry because, unlike the larger fallout particles, it can be inhaled. James R. McCarroll, EPRI's health effects research program manager, calls acid mist "the single most important health question in electricity research." Yet he admits that "little, if any, work is being done on health effects from fallout."

Physical effects. Property damage from the larger particles is far better documented, and the CARB estimates damage at $6.25 million in the Redondo Beach area alone. Throughout California, utilities have been reimbursing property owners for damage in hope of avoiding a regulatory crackdown. Still, Southern California Edison has already been slapped with two class-action suits, and local environmental agencies have ordered Southern California utilities to halt the acid smut emissions.

Ironically, there is ample evidence that the problem can be solved. East Coast utilities have for years used fuel additives to curb acid formation in order to protect their boilers. Most of the techniques have not gotten rid of enough acid to prevent fallout, however, and many government regulators are bitter about what they see as the utilities' unwillingness to pursue controls more strenuous-

ly. "There's no doubt in my mind that industry has given acidic formation the fast shuffle," complains James J. Morgester, the CARB's chief of enforcement.

Understandably, experts within the industry are unwilling to agree publicly with this. Privately, though, many concede that the lack of federal regulations governing acid smut and the all-out effort to study acid mist instead have caused the problem to receive short shrift. Still, a few utilities have moved on their own, using fuel additives and modified air-to-fuel ratios to retard acid formation.

Success stories. Boston's Stone & Webster Engineering Corp. has worked out successful revised combustion methods for New England Power Co. and a few other East Coast utilities. And San Diego Gas & Electric Co. has successfully tested a magnesium-based additive since June, 1976. Company spokesmen say that they started testing the additive, manufactured by Rolfite Co., in a second power plant last month and that acid fallout has dropped almost 90%.

Perhaps Lilco's is the best success story. Almost 13 years ago, the utility started using less excess air in combustion, as well as a magnesium fuel additive, and the result was a tenfold drop in sulfur trioxide concentrations—from 60 parts per million to 6 ppm. At the same time, Lilco found that burning magnesium oxide and fuel oil yielded vanadium, an element prevalent in the Venezuelan oil that the company uses. Last year, Lilco sold 362 tons of recovered vanadium—9% of the nation's total requirements—for $1.2 million. Other annual savings: $2 million in fuel because of thermal efficiencies and $400,000 from reduced boiler corrosion.

"We spent $600,000 for magnesium oxide and benefited by $4 million," O'Neal crows. Even more important, he says, the new system stopped the daily complaints about property damage that the company had previously received. "Complaints are down 90%," he says, "and that's a benefit that can't be measured in dollars and cents." ∎

opportunity to satisfy growing demands through actual expansion of the utilities' markets. On the other hand, universal use of solar energy for heating and cooling would severely curtail the market for conventional energy on sunny days, and pose severe peak-demand problems on cloudy days. Combined solar and conventional energy for heating and cooling is not only advantageous, but enables the utilities to have their cake and eat it too: They keep their existing market for conventional energy and expand their total market through the control of solar energy.

To some, this may sound ominous. To others who know that utilities understand the high start-up costs and long-term amortization of capital-intensive business (as opposed to the homeowner who may only look at first-cost, and not the life-cycle costs), the idea makes sense.

We don't know who is right; only that you, the homeowner contemplating the installation of a solar heating system, should be conscious of the possible legal interactions between your state, your utility company, and your solar future.

Labor Laws

We don't envision many labor problems occurring during these early days of solar heating, but when the pace of the activity picks up, you can look for headaches. The conflicts will occur in two arenas; first, jurisdictional problems between carpenters, sheet metal workers, plumbers and others, each wanting their fair share of the solar equipment installation pie. The other arena of conflict will be that of the factory prefabricated subassembly proponents versus the do-it-at-the-site ones.

The whole matter is complicated by the fact that the National Labor Relation's Act preempts most state legislative action which could pave the way for harmony and productive labor peace. So, we—the authors—can do nothing more at the moment than alert you to this possible obstacle, and suggest that you move fast, and complete your solar installation before the top brass of each trade union begins to think that some other craft may be getting its share of the pie.

Insurance

It will still be awhile before all insurance companies have evaluated the risks involved with solar equipment, and the rates they must charge to insure against those risks. Many passive and active solar collectors are covered with sheets of glass, and make a tempting target for small children with stones. On the other hand, suppose a motorist is suddenly blinded by the glare of a nearby solar collector, and has a serious and costly accident. Who is liable for the damage? What if the additional weight of the collectors on a roof cause it to sag or collapse? Is there insurance coverage available for that possibility?

These and many similar questions still need answers. We know that good design and careful construction can greatly reduce the risks that might otherwise be present. But is good quality work good enough? For some of you, it will be; for others, the thought of proceeding without insurance will be unacceptable. If you're part of the latter group, contact your homeowner's insurer and discuss the matter with him. You'll probably find that coverage is available at a slightly increased premium.

Some other subjects which could provide provocative food for thought on the subject of solar/legal considerations are—

- Patents and their restrictions to the free use of certain solar technology.
- Banking law and its ways of inducing (or impeding) solar development. We'll be discussing this further in the next chapter, on financing your solar system.
- Fiscal policies, such as direct government bonuses or subsidies.
- Definitions. For instance, if the government did offer a subsidy for installing solar-collecting devices, would your new swimming pool qualify? It certainly does collect the sun's energy when the water in the pool warms.

If you now have a new perspective on solar energy's legal considerations, we've accomplished our objective. Solar heating will come of age more quickly, and in a more dignified way, if those who buy it are well informed. Don't be frightened or intimidated, but *do be aware.*

A Summary of State Actions and Incentives Supporting Solar Usage

STATE	BILL #	DESCRIPTION	CONTACT
Arizona	S.B. 1011 (1975)	Provides for amortization of the cost of solar energy devices over 60 months in computing net income for state income tax purposes. Applies to all types of buildings.	Neal Trasente, Director Dept. of Revenue Capitol Building, West Wing Phoenix, AZ 85017 (602) 271-3393
	S.B. 1231 (1974)	Provides exemption from property tax increases which may result from addition of solar system to new or existing housing.	same as above
California	S.B. 218 (1976)	Provides for state income tax credit (up to $1,000) of 10% of the cost of installing a solar system.	
Colorado	S.B. 75 (1975)	Provides that all solar systems be assessed at 5% of their original value for property tax purposes.	Raymond E. Carper Property Tax Administrator Col. Division of Property Taxation 614 Capitol Annex Denver, CO 80203 (303) 892-2371
	S.B. 95 (1975)	Provides procedures for recording voluntary solar easements.	Relevant County Clerk and County Recorder
Connecticut	S.B. 652 (1976)	Provides that solar system shall not increase property tax assessment of real property on which it is located. Subject to authorization by local ordinance.	Relevant Local Assessor or Board of Assessors

STATE	BILL #	DESCRIPTION	CONTACT
Georgia	H.B. 1480 (1976)	Provides that purchasers of equipment for solar systems will receive refund of sales and use tax paid on such equipment.	Dept. of Revenue Sales Tax Unit Room 310 Trinity—Washington Bldg. Atlanta, GA 30334
	Constitutional Amendment	Authorizes governing authority of any county or municipality to exempt value of solar system from ad valorem property taxation.	Relevant city council or county board of supervisors
Hawaii	S.B. 2467	Provides income tax credit for up to 10% of cost of solar system for year of installation. Also provides exemption from any property tax increase resulting from addition of solar system.	Tax Department Hale Auhdo Building 425 Queen Street Honolulu, HI 96813
Idaho	H.B. 468	Provides that cost of residential solar system may be deducted from taxable income over a period of four years. Deduction shall not exceed $5,000 in any one taxable year.	Income Tax Division State Tax Commission P.O. Box 36 Boise, ID 83722 (208) 384-3290
Illinois	H.B. 164 (1975)	Provides that when a solar system has been installed on real property, owner of property may claim improvement value of a conventional system if that value is less than value of solar system.	Frank A. Kirk, Director Dept. of Local Government Affairs 303 East Monroe St. Springfield, IL 62706 (217) 782-6436
Indiana	S.B. 223 (1974)	Allows owner of real property with solar system an annual deduction from the assessed value of the property equal to the lesser of: (1) the assessed value of the property with solar system minus the assessed value without the system, or (2) $2,000.	County Auditor

STATE	BILL #	DESCRIPTION	CONTACT
Kansas	H.B. 2969 (1976)	Allows individuals and businesses to deduct 25% of cost of solar system (up to $1,000 for individuals and $3,000 for businesses) from state income tax.	Kent Kalb, Secretary of Revenue
Maryland	H.B. 1604 (1975)	Provides that solar system shall not cause property tax assessment of new or existing building to be greater than it would be with conventional system.	Albert W. Ward, Admin. Dept. of Assessment and Taxation 301 W. Preston Street Baltimore, MD 21201 (301) 383-2526
Massachusetts	S.B. 1664 (1976)	Provides that a corporation may deduct cost of a solar system from its taxable income for year of installation. Also provides that solar system will not be subject to tangible property tax.	Harvey Michaels Asst. to the Director for Solar Energy Energy Police Office John McCormack State Office Building One Ashburton Place Boston, MA 02108 (617) 727-4732
	H.B. 6813 (1975)	Provides for real estate tax exemption for solar system.	Harvey Michaels Asst. to the Director for Solar Energy Energy Police Office John McCormack State Office Building One Ashburton Place Boston, MA 02108 (617) 727-4732
Michigan	H.B. 4137 (1976)	Provides that receipts from sale of tangible property to be used in solar system shall not be used to compute tax liability for business activities tax.	Edward Kane, Director Dept. of Treasury State Tax Commission State Capitol Bldg. Lansing, MI 48922

STATE	BILL #	DESCRIPTION	CONTACT
Michigan (con't.)	H.B. 4138 (1976)	Provides that tangible personal property used for solar devices shall be exempt from excise tax on personal property.	same as above
	H.B. 4139 (1976)	Provides that solar system shall not be considered in assessing value of real property.	same as above
Montana	H.B. 663 (1975)	Encourages investment in nonfossil forms of energy generation and in energy conservation in buildings through tax incentives and capital availability.	William Groff, Director Dept. of Revenue Michell Building Helena, MO 59601 (406) 449-2460
New Hampshire	H.B. 479 (1975)	Allows each city and town to adopt (by local referendum) property tax exemptions for solar systems.	Local or municipal tax assessor
New Mexico	S.B. 1 (1975)	Provides for tax credit against state personal income tax for 25% of cost of solar system up to $1000.	Marshal Morton Bureau of Revenue Manuel Lujan, SR, Bldg. St. Francis Drive at Alta Vista Sante Fe, NM 87503 (505) 827-3221
North Dakota	S.B. 2439 (1975)	Provides that solar heating and cooling systems utilized in new or existing buildings will be exempt from property taxes for five years following installation.	John Hulteen Supervisor of Assessments State Tax Department State Capitol Bismarck, ND 58505 (701) 224-3461

STATE	BILL #	DESCRIPTION	CONTACT
Oregon	H.B. 2202 (1975)	Provides exemptions from *ad valorem* taxation for any increased value of property resulting from installation and use of a solar system.	Donald M. Fisher, Admin. Assessment and Appraisal Division 506 State Office Bldg. Salem, OR 97310 (503) 378-3378
	H.B. 2036	Adds solar energy considerations to comprehensive planning. Allows city and county planning commissioners to recommend ordinances governing building height for solar purposes.	Lon Topaz, Director Dept. of Energy 528 Cottage Street, N.E. Salem, OR 97310 (503) 378-4128
South Dakota	S.B. 283 (1975)	Allows owner of residential real property an annual deduction from assessed value for installation of solar device. Deduction may be equal to the lesser of: (1) difference between assessed value with system and value without, or (2) $2,000.	County Auditors/Assessors and Lyle Wendell, Sec. Dept. of Revenue State Capitol Pierre, SD 57501
Texas	H.B. 546 (1975)	Provides exemption from sales taxes on receipts from sale, lease or rental of solar devices. Business tax exemption granted to corporations which exclusively manufacture, install or sell solar devices. Corporation may deduct amortized (60 months) cost of solar system from taxable capital.	Bob Bullock, Comptroller Director of Public Accounts LBJ Building 17th & Congress Austin, TX 78711
Vermont	H.B. 206	Allows towns to enact real and personal property tax exemptions for solar systems.	State Energy Office State Office Building Montpelier, VT 05602 (802) 828-2768
Virginia	Constitutional Amendment	Permits tax exemptions on property used for solar energy systems.	Relevant city council or county board of supervisors

The following two state laws are included here as samples to give you some idea of how a new law—or change to an existing law—reads.

Illinois State Law Supporting Solar Usage

AN ACT to add Sections 20d-1, 20d-2 and 20d-3 to the "Revenue Act of 1939," filed May 17, 1939, as amended.

Be it enacted by the People of the State of Illinois, represented in the General Assembly:

Section 1. Sections 20d-1, 20d-2 and 20d-3 are added to the "Revenue Act of 1939," filed May 17, 1939, as amended, the added Sections to read as follows:
(Ch. 120, new par. 501d-1)
Sec. 20d-1. It is declared to be the policy of the State of Illinois that the use of solar energy heating or cooling systems should be encouraged as conserving nonrenewable resources, reducing pollution and promoting the health and well-being of the people of this State, and should be valued in relation to these benefits to the people of the State.
(Ch. 120, new par. 501d-2)
Sec. 20d-2. "Solar energy heating or cooling system" means any system, method, construction, device or appliance designed, constructed and installed relying on the use of the sun's rays rather than on conventional heating or air conditioning systems, for heating or cooling a building, which conforms to the standards for such systems established by regulation of the Department of Local Government Affairs, or its successor agency.
(Ch. 120, new par. 501d-3)
Sec. 20d-3. When a solar energy heating or cooling system has been installed in improvements on any real property, the owner of that real property is entitled to claim an alternate valuation of those improvements. The claim shall be made by filing with the county assessor, supervisor of assessments or board of assessors, as the case may be, a certified statement, on forms prescribed by the Department of Local Government Affairs, or its successor agency, and furnished by the assessing officer, setting out (a) that the specified improvements on described real estate are equipped with such a system, (b) that the system is used for heating or cooling or both heating and cooling those improvements, and (c) the total cost of the solar energy heating or cooling system.

When such a statement and claim for alternate valuation is filed, the county assessor, supervisor of assessments or board of assessors, as the case may be, shall ascertain the value of the improvements as if equipped with a conventional heating or cooling system and the value of the improvements as equipped with the solar energy or cooling system. So long as the solar heating or cooling system is used as the means of heating or cooling those improvements, the alternate valuation computed as the lesser of the two values ascertained under this paragraph shall be applied. Whenever the solar heating or cooling system so valued ceases to be used as the means of heating or cooling those improvements, the owner of that real property shall within 30 days notify in writing by certified mail, return receipt requested, the county assessor, supervisor of assessments or board of assessors, as the case may be, of that fact. It shall be a Class B misdemeanor to fail to submit information required under this Section or to knowingly submit any false information required under this Section.

Indiana State Law Supporting Solar Usage

AN ACT to amend IC 1971, 6-1 by adding a new chapter concerning property tax deductions for solar energy systems.

Be it enacted by the General Assembly of the State of Indiana:

SECTION 1. IC 1971, 6-1 is amended by adding a new chapter 9.5 to read as follows:

Chapter 9.5 Deduction-Solar Energy Systems.

Sec. 1. The owner of real property which is equipped with a solar energy heating or cooling system may have deducted annually from the assessed valuation of the real property a sum which is equal to the lesser of:

(1) the remainder of (i) the assessed valuation of the real property with the solar heating or cooling system included, minus (ii) the assessed valuation of the real property without the system; or

(2) two thousand dollars ($2,000).

Sec. 2. The owner of real property who desires to claim the deduction provided in this chapter must file a certified statement in duplicate with the auditor of the county in which the real property is located. In addition, the owner must file the statement on forms prescribed by the state board of tax commissioners, and he must file the statement between March 1 and May 10, inclusive, of each year for which he desires to obtain the deduction. Upon verification of the statement by the assessor of the township in which the real property is located, the county auditor shall make the deduction.

BIBLIOGRAPHY

Business Law. R. A. Anderson and W. A. Kumpf, 5th edition, Cincinnati: South-Western Publishing Co.

"Law and Solar Energy Systems: Legal Impediments and Inducements to Solar Energy Systems," Richard L. Robbins, *Solar Energy* 18:371–79, Elmsford, NY 10523: Pergamon Press, 1976.

"Local Florida Planners Kill Solar Collector Ban," *Solar Utilization News*, vol 2, no. 1, July 1977, Alternate Energy Institute, P.O. Box 3100, Estes Park, CO 80517.

Proceedings of the Workshop on Solar Energy and the Law. William A. Thomas (principal investigator), American Bar Foundation, March 1975. Available from National Technical Information Service, Springfield, VA 22161; order number PB-241-051.

The Solar Conspiracy. John Keyes. Dobbs Ferry, NY 10522: Morgan & Morgan, 1975.

Sunspots. Steve Baer, Zomeworks Corp., Box 712, Albuquerque, NM 87103, 1975.

A Survey of State Legislation Relating to Solar Energy. Robert M. Eisenhard, U.S. Department of Commerce, National Bureau of Standards, Washington, DC 20234. Available from National Technical Information Service, Springfield, VA 22161; order number PB–258–235, April 1976.

"Who Will Own the Sun?" Mark Northcross, *The Progressive*, April 1976, pp. 14–16.

9

The Sun Isn't Free

If you're one of those fortunate persons who can dip into his piggy bank for the cost of a solar system please feel free to skip past this entire chapter, but if, like most of us, you have to borrow money for such a project, a few words about loans might be in order.

It's probably best to warn you that you may get a disappointing reaction when you walk up to the loan officer at your bank or mortgage company, and announce that you want to borrow money for a solar heating system. Loan officers are, of course, wise and wonderful people, but, absorbed as they are in their interest tables and credit ratings, they sometimes fail to notice what's happening outside . . . in the real world. The dramatic rise in the price of all fuels is well known to them, but the related fact—that solar energy can sharply reduce home heating bills—will probably be another year or two in capturing their attention. Until then, many a homeowner applying for a solar heating loan will continue to be asked what in the world he's talking about: "Solar heating? What's *that*?"

Prepare yourself for a disappointment. The fact that you are about to embark upon a satisfying, money-saving, home-improving, and patriotic adventure is likely to be lost upon your banker, so you might as well request a loan to cover the cost of a new heating system, and let it go at that. If he or she asks for further details on the system you propose to buy, that's the time to start your solar lecture.

THE PHILOSOPHY OF BORROWING

At one time or another each of us has asked to borrow something. Perhaps it's been nothing more than a garden tool from a neighbor, or a cup of milk when we've been caught short. No matter what, we probably made our loan request in a somewhat apologetic manner. We were asking a favor from someone who was in the position of granting or not granting it, and it put us in an uncomfortable position.

This apologetic manner, when asking someone to favor you with a loan, usually becomes more intense when you approach a lending institution and ask for money. Well-established institutions in buildings dating back to pre-World War II days are intimidating in appearance;

Well-established institutions are intimidating in appearance.

oftentimes their massive facades make a person feel small and insignificant. Then there is the long and sometimes embarrassing wait while the loan officer, behind his sacrosanct marble and brass barrier, shuffles papers and pretends you're not there waiting to petition him for his generosity. And finally, when you *are* ushered into the presence of the lender, his cool stare and superior attitude reduce you to a quivering mass of jelly. You try to state your case, but the story comes out disjointed and unorganized, like that of a third-grade student trying to tell his teacher his perfectly good reason for being late. You feel like an idiot, and as you and your loan request seem to diminish in importance, the loan officer of the great institution seems to grow more God-like in stature.

A familiar scene? It's been reenacted on television and in the movies hundreds of times, and many of us have experienced it personally at one time or another.

It shouldn't be that way! It's built on a myth . . . a fairy tale we've been exposed to for so long we've actually come to believe it. It's propaganda, and the lenders *want us* to accept it that way; it makes their job so much easier.

The *fact* is that you are granting them the favor! "It doesn't make sense," you say. "How can I be granting the favor when I'm the one asking for the loan?" Simple. You have been conditioned to think that you are *asking to borrow,* but you are really *offering to purchase* the only commodity the lending institutions have for sale. What do they sell? The use of their funds. Just as the merchant sells his wares to earn a living, and the lawyer his counsel, so too, do the lending businesses sell their stock-in-trade: money.

For example, a savings and loan association may have a million dollars in available funds to be lent to qualified borrowers. For the use of these funds, the association pays its depositors perhaps 7% a year, or

The loan officer.

$70,000. When those funds are "sold" to borrowers, the going yearly rate might be 9¾%, which returns $97,500 to the savings and loan association annually. The difference between the "cost" of the funds and the "sales price" of the funds ($97,500–$70,000) is $27,500. This is the margin from which the lending institution pays its operating expense and returns a profit to its owners.

Just as the ice cream shop cannot operate with only one can of ice cream, these lending institutions cannot make money by lending only a little. Some have many millions (and sometimes billions) of dollars available for "sale." In order to avoid failure, and perhaps bankruptcy, lenders must keep their "merchandise" moving. What they get for their funds must be greater than what they pay for their funds . . . or they're in trouble. Their job, then, is to lend ("sell") *their available funds at a profit to qualified borrowers.* Each day that the available dollars sit in the vault, unused, they are losing money (as interest payments to their depositors), so you *really are granting lenders a favor* when you apply for a loan.

To more fully understand the philosophy of borrowing, you must be conscious of the fact that lending institutions compete with one another, just as Gulf Oil does with Exxon. When you're shopping for a new appliance, you check for price and quality everywhere—in catalogues, department stores, specialty shops, newspaper ads, etc.—to get the best return for your purchase price. You can, and should, do the same when you shop for a cash loan. Don't be afraid to let the lender know that you're looking for the best "buy"; if you do, he'll recognize you as a knowledgeable borrower. Remember . . . if you're a qualified borrower, you are in a position to favor the lender with your business.

Later in this chapter, when we review *lender's concerns,* we will talk about what constitutes the "qualified" borrower. For now, though, let's review options available to you in your quest for funds.

THE BORROWER'S OPTIONS

When borrowing for residential purposes, there are three basic ways of obtaining funds. They are—

1. first mortgage loans (conventional, FHA, VA, FmHA)
2. second mortgage loans
3. home improvement loans (Title I and conventional)

Your own circumstances should help decide the type of loan you try for first. Certainly you want the least expensive that will answer your needs. But because there are so many variables to consider, we'll have to explore each one and let you judge for yourself which type of loan would be the most suitable for your specific requirements.

Remember, we're about to discuss *types of loans,* not *types of lending agencies.* They're two different subjects. Cost-wise, first mortgage loans are normally less expensive than home improvement loans, which, in turn, are usually less expensive than second mortgage loans. The table here will give you some insight into the relative cost of the three types.

HOW FINANCING AFFECTS SOLAR COSTS
Illustrative Loan Terms and Monthly Debt Service
Under Private Lender Financing Alternatives

| | LOAN TYPE | | | | | | | |
| | FIRST MORTGAGE | | | | | SECOND MORT-GAGE | HOME IMPROVE-MENT | |
	Conventional			FHA	VA	Con-ventional	Title I	Con-ventional
Loan/Value Ratio	70%	80%	90%	93%	100%	75%	100%	100%
Interest Rate	8.5%	8.75%	9.0%	8.25%	9.0%	13.5%	11.5%	12.5%
Maturity (years)	27	27	27	30	30	10	12	5
Mortgage Insurance	—	.15%	.25%	.5%	—	—	.5%	—
Monthly Cost per $1,000 of Loan	$7.88	$8.16	$8.41	$8.06	$8.23	$15.23	$13.13	$22.50

Source: HUD-PDR-218, Home Mortgage Lending and Solar Energy, published March 1977.

Quite a spread isn't it? From $7.88 to $22.50 per thousand borrowed. You can see that conventional mortgages are the least expensive, so why doesn't everyone use them?

To start with, first mortgages are primarily for *new* residential construction or for the purchase of an existing residence. If you need financing for a solar installation on your *present* home, the only mortgage route you can take is to remortgage your home. Most often this is unwise, since the rate of your existing home mortgage is probably lower than that of any new mortgage you could negotiate. The lending institutions would love to have you pay off your 7% mortgage with the proceeds of a new 9¾% mortgage. (The new one would include the amount being borrowed for the solar installation.) It might also make the monthly cost at first appear lower since the loan would be paid back over a twenty- to thirty-year period, but the overall cost would make this an unwise choice for any solar retrofit financing.

NEW HOMES

On the other hand, a first mortgage *is* the most desirable way of financing a totally new, solar-heated residence. Depending on the amount of down payment you can afford, several types of first mortgages are available for your new home loan:

- For relatively low-income farm families, there is the FmHA loan (Farmer's Home Loan Agency). An FmHA loan has many restrictions, but is by far the least expensive (the loans are sometimes as low as 1 or 2%) and it requires little by way of down payment. If your family income is $5,000 or less (for a family of four) be sure to investigate this possibility with your local FmHA office.
- VA (Veteran's Administration) guaranteed loans also require minimal down payment and cost much less than home improvement loans or second mortgages. The problem here is that the VA does not actually make the loan; it only guarantees to the lender that if you fail to make your payments, the VA will assume your obligation, and make the payments to the lender (while trying to sell your home to recover the loss). Because of this, you have the two-step task of getting VA approval for your loan *and* finding a lending institution that will give you the loan under the VA guidelines.
- FHA (Federal Housing Administration) loans require that you have at least 7% of the loan as a down payment. On a $50,000 home, this means you must be prepared to make a down payment of at least $3,500. Again, the FHA merely guarantees your loan, and you face the two-step loan procedure we just mentioned in connection with VA loans. For new solar homes, though, the FHA guaranteed loan is the best route to take when you're limited to a small down payment, and don't wish to do too much shopping for competitive rates.
- Conventional mortgages require greater down payments, generally at least 20% of the cost of the new home ($10,000 on that

$50,000 purchase), but here you enter the truly competitive market. Many dollars can be saved by locating the bank or lender that has surplus funds and is eager to find a qualified borrower with whom those surplus dollars can be put to work.

RETROFIT SYSTEMS

But suppose you have an existing home, and are not interested in remortgaging the whole structure for your solar installation loan. What other options are available? The most common solution is the *home improvement loan*. Under this heading comes a variety of lending arrangements offered by every kind of lending company.

• The *conventional home improvement* loan is generally available from savings and loan associations, savings banks, commercial banks, credit unions, and even private finance companies. The cost ranges anywhere from about 12.5% a year up. Generally, it is limited to a five-year maturity, so monthly payments are higher than they might be with a longer-life mortgage loan.

• A *Title I home improvement* loan refers to home modernization loans made under Title I of the Federal Housing Act. Its significance lies in the fact that its maturity can extend for as much as twelve years, thereby greatly reducing the monthly payment on this government-backed loan. Its cost, at about 11.5% (mid 1977), is also slightly less than that of the conventional home improvement loan.

• And finally, at a slightly higher cost than that of the conventional home improvement loan, is the *second mortgage*. It is similar to a first mortgage except that the holder of the second mortgage—that is to say, the lender—does not have the first claim to your property if you fail to repay on your loan. The *first* mortgage holder must be satisfied before the *second* mortgage holder can exercise his legal rights. Obviously, then, this is a "last resort" position for your solar installation loan, for although the payback period can extend to ten years, the cost will be from 1 to 2% higher than that of other home improvement loans because of the added risk to the lender.

THE LENDER'S CONCERNS

Before we examine the lender's concerns with regard to your proposed new solar installation, we'll first look at you the borrower from the lender's point of view. "Who is an 'easy' loan applicant?" This question, in a recent survey of lending officers,* generated a typical response: "A married, young, affluent engineer . . . looking for 40% financing on an $80,000 home that he plans to live in for a long time."

Let's dissect that statement to see what those typical loan officers are really looking for:

Statement	Means
"A married . . .	The borrower should be a responsible member of the establishment.
young . . .	He (or she) should have many working and earning years ahead.
affluent . . .	His present financial status should be solid, so that unexpected cash needs won't cause serious problems.
engineer . . .	His profession should be one for which a continued need is probable.
looking for 40% financing . . .	His financial position should be so strong that he can afford a 60% ($48,000) down payment on his new $80,000 home.
on an $80,000 home . . .	A substantial home, in a good neighborhood, which is sure to increase in value as years pass by.
that he plans to live in for a long time."	The longer a borrower plans to stay in his home, the more money he will be repaying, and therefore the more secure the balance due to the lender.

It's a shame, isn't it? And yet if you were the lender, you too would want the "sure bet" for repayment, in order to minimize any potential default problems. Certainly we're not all married, young, affluent engineers . . . but you get the idea of what the lenders would like—*if* they had the choice.

The following home improvement loan application is fairly typical, and if you read "between the lines" with us, you'll see quite clearly what the lender really wants to know.

*Survey conducted in early 1976 by Regional and Urban Planning Implementation, Inc. (RUPI) with support from the National Science Foundation.

APPLICATION FOR CREDIT TO THE BANK OF

PLEASE PRINT AND COMPLETE IN FULL. INDICATE AMOUNT AND PURPOSE ON REVERSE SIDE

YOUR PERSONAL HISTORY

	FIRST	MIDDLE INITIAL	LAST	AGE	MONTH	DAY	YEAR	SOCIAL SECURITY NUMBER
NAME (PRINT)				DATE OF BIRTH				

	NUMBER	STREET	CITY	STATE	ZIP CODE	HOW LONG?
HOME ADDRESS (PRINT)						
PREVIOUS HOME ADDRESS	NUMBER	STREET	CITY	STATE	ZIP CODE	HOW LONG?
PREVIOUS HOME ADDRESS	NUMBER	STREET	CITY	STATE	ZIP CODE	HOW LONG?

(GIVE 5 YEARS CONTINUOUS ADDRESSES—USE REVERSE SIDE IF NEEDED)

HOME PHONE	BUSINESS PHONE	NUMBER OF DEPENDENTS	DO YOU OWN HOME? YES ☐ NO ☐	IF YES PURCHASE PRICE $	IF YES MARKET VALUE $

AUTO OWNED	YEAR	MAKE	OWN OTHER REAL ESTATE WHERE?	OTHER ASSETS (DESCRIBE)

NAME AND ADDRESS OF RELATIVE NOT LIVING WITH YOU

YOUR INCOME

	NAME OF EMPLOYER	ADDRESS		HOW LONG?
ARE YOU SELF EMPLOYED? ☐ YES ☐ NO	OR YOUR TRADE NAME			

				BANK	BRANCH
POSITION OR OCCUPATION	BADGE OR CLOCK NUMBER	INCOME $	PER	BUSINESS BANK ACCT. IF SELF EMPLOYED	

	ADDRESS			HOW LONG?
PREVIOUS EMPLOYER		POSITION		

(GIVE 5 YEARS CONTINUOUS EMPLOYMENT—USE REVERSE SIDE IF NEEDED)

OTHER INCOME—Please Note: You are not required to reveal receipt of alimony, child support or maintenance payments in connection with this application. You may at your option describe any alimony, child support or maintenance payments or other regular income (taxable or not) you wish considered in evaluating your application.

AMOUNT OF OTHER INCOME $	PER	SOURCE OF OTHER INCOME

SPOUSE

INFORMATION ABOUT YOUR SPOUSE—CO-APPLICANT

Please Note: You are not required to furnish any information requested about your spouse or former spouse, as applicable, unless your spouse wishes to sign the application as a co-applicant and be contractually liable on the account, or you are relying upon alimony, child support, maintenance payments or the income from a spouse or former spouse as a basis for repayment of the credit you are applying for.

		AGE	MONTH	DAY	YEAR	SOCIAL SECURITY NUMBER
SPOUSE'S FULL NAME		DATE OF BIRTH				

		HOME PHONE	BUSINESS PHONE
SPOUSE'S ADDRESS IF DIFFERENT FROM ABOVE			

	NAME	ADDRESS	OCCUPATION OR POSITION	INCOME $	PER
SPOUSE'S EMPLOYER					

FINANCIAL INFORMATION

LIST ALL DEBTS INCLUDING THIS BANK - AUTO - FINANCE CO. LOANS - DEP'T. STORES - IF NO DEBTS STATE "NONE". USE OTHER SIDE IF NEEDED.

NAME OF CREDITOR	ADDRESS	ACCOUNT NUMBER	BALANCE	MONTHLY PAYMENT
MORTGAGE HELD BY			$	$
NAME OF LANDLORD		XXX XX	$ XXX XX	$
PAYING ALIMONY OR CHILD SUPPORT TO		XXX XX	$ XXX XX	$
AUTOMOBILE FINANCED BY			$	$
OTHER DEBTS (LIST HERE AND BELOW)			$	$
			$	$
			$	$
			$	$

ARE ANY OF THESE DEBTS PAST DUE IF YES—EXPLAIN—(ATTACH SHEET)	I HEREBY REPRESENT THAT MY TOTAL INDEBTEDNESS DOES NOT EXCEED	$

BANKS

TYPE ACCOUNT	BANK NAME	ADDRESS	BRANCH	ACCOUNT NUMBER	BALANCE
CHECKING					
SAVINGS					

REFERENCE

LIST CREDIT CARDS YOU HOLD:

HAVE YOU HAD ANY LOANS AT THIS BANK IN THE PAST SEVEN (7) YEARS? ☐ YES ☐ NO	☐ OPEN	☐ PAID IN FULL	ACCOUNT NUMBERS (IF AVAILABLE)

DO YOU HAVE ANY LOAN APPLICATIONS PENDING AT ANY FINANCIAL INSTITUTION AT THIS TIME? ☐ YES ☐ NO (IF YES—EXPLAIN)

HAVE YOU NOW—OR EVER HAD—ANY JUDGMENTS GARNISHMENTS OR LEGAL PROCEEDINGS AGAINST YOU? ☐ YES ☐ NO (IF YES—EXPLAIN)

PLEASE HAVE PAYMENTS FALL DUE ON 5TH ☐ 10TH ☐ 15TH ☐ 20TH ☐ 25TH ☐ OF MONTH. (MINIMUM 20 DAYS—MAXIMUM 50 DAYS FROM DATE OF NOTE)

SIGNATURE OF CO-APPLICANT	DATE	SIGNATURE OF APPLICANT

AMOUNT OF LOAN REQUESTED—$ _____ **FOR** _____ **MONTHS FOR THE FOLLOWING PURPOSE(S)**
NUMBER

PLEASE CHECK APPROPRIATE BOX(ES) BELOW AND GIVE INFORMATION REQUESTED:

☐ PERSONAL (Describe Purpose) _____

_____

☐ TO BUY: ☐ AUTO ☐ BOAT ☐ OTHER
         ☐ TRUCK ☐ CAMPER (Describe) _____

NEW OR USED	YEAR	MAKE OR TRADE NAME	MODEL IF TRUCK GIVE G.V.W. IF BOAT GIVE TYPE & LENGTH	NO. CYLINDERS OR NO. ENGINES	SERIAL NUMBER	LICENSE NUMBER	KEY NUMBER

EQUIPPED WITH ITEMS CHECKED
☐ Radio   ☐ Automatic Trans.   ☐ Power Steering   ☐ Power Windows   ☐ High Performance Engine—Cu. In. Disp._____H.P._____
☐ Tinted Glass   ☐ 4 Speed Trans.   ☐ Power Brakes   ☐ Power Seats   ☐ Air Conditioning

		DEALER OR SELLER'S NAME		BANK USE
CASH PRICE	$		INV.	
CASH DOWN PAYMENT	$	DEALER OR SELLER'S ADDRESS	L.	/H
TRADE-IN ALLOWANCE	$	INSURANCE AGENT	L.V.	
AMOUNT TO BE FINANCED	$	INSURANCE AGENT ADDRESS		

☐ HOME IMPROVEMENT TO: ADDRESS OF PROPERTY TO BE IMPROVED

PROPERTY IN NAME OF:	☐ ONE FAMILY ☐ MULTIPLE FAMILY	MARKET VALUE	ANNUAL TAXES	MORTGAGE BALANCE
	☐ TWO FAMILY ☐ OTHER	$	$	$

NAME OF CONTRACTOR	ADDRESS OF CONTRACTOR	CONTRACT TOTAL COST	$
		DOWN PAYMENT	$
		AMOUNT TO BE FINANCED	$

NOTICE—It is understood that this bank does not recommend or approve any contractor, dealer, or salesman. The selection of and negotiations between the borrower and any contractor, dealer, or salesman is solely the responsibility of the borrower. Any subsequent sale or transfer of the property for which this loan was made, automatically requires payment of loan balance in full.

APPLICANT OR CO-APPLICANT USE THIS SPACE FOR ADDITIONAL INFORMATION. _____

_____

_____

_____

_____

THIS SPACE FOR BANK USE ONLY

1. Your personal history—how old you are, how stable; what property do you own; what other assets?

2. Your work history—again how stable and responsible you are, and have you any other income?

3. Your marital history—again how stable you are, and what other income is your spouse bringing into the family?

4. Your debt history—can you afford another financial obligation without being overburdened?

5. Your banking history—and again, how stable are you, and how affluent (based on your savings account)?

6. Your loan history—have you borrowed before, and how timely have your repayments been?

7. What's the loan for?—As we said on the first page of this chapter, a statement to the effect that the home improvement loan is to cover the cost of a new heating system should be adequate (*if* your other responses indicate that you're a good credit risk).

By now you should have a good idea of what to say, and what not to say, in a loan application. We certainly don't suggest that you misstate facts in your application (that could be a serious mistake, and may be a crime), but we do feel that by exercising care in what is and is not stated, you can enhance the image you present to the lender as a "desirable" borrower. And don't forget for one moment, that if you are a good credit risk, you are doing the lender a favor by bringing your business to him! We'd like to emphasize this point with a brief story that took place a few years ago.

A thirty-five-year-old accountant, who had worked with a major industrial firm in one location for more than fifteen years, went to his local bank for a home improvement loan of $2,500. He planned to install new storm windows, doors, and insulation on his home, where he, his wife, and two daughters had lived for more than twelve years. Perhaps because his skin was black, or perhaps because he had never before borrowed from the bank, the accountant was told by the lending officer that a cosigner would be required before the loan could be granted.

After we heard this story and saw what a good credit risk the accountant was, we suggested he inform the local banker that the business would be taken elsewhere if his background and signature were not considered adequate for the loan.

Convinced that the banker would quickly show him the way out, he nonetheless followed our advice. After listening to our friend's threat

to take his business elsewhere, the lender quickly conferred with his
superior at the bank. Both men then returned to our friend, and with
apologies, informed him that his loan application would be quickly
processed . . . on the basis of his signature alone.

So you see, it's true: The lenders really do need qualified borrowers,
as much (if not more) than you need the lenders.

What about lender concerns that specifically relate to solar installations? As we said at the beginning of this chapter, many lenders may
still ask, "Solar heating? What's that? But this won't be the general rule.
Most lenders are personally conscious of *and* *supportive* of the use of
solar energy, but their personal attitudes will not affect their business
judgment when it comes to granting loans. For now, and in the near future, there are three major areas of uncertainty that the lender must
evaluate:

1. the impact of solar installations on property value
2. the technical performance of the systems
3. the estimated future savings in energy costs

In future years the impact of solar installations upon property
values will be known, understood, and obvious. This is not necessarily
so now, though. In fact, it's possible that the cost of a solar installation
may not be fully reflected by a higher sales price for a home, if it were to
be resold. Lenders are very cautious of possible "overimprovements"
such as copper flashing, marble floors, three-car garages, and swimming
pools. In some neighborhoods, a swimming pool can add as much value
to a house as the pool initially cost, while at another house it would not.

And so it is with solar installations; the lenders really don't know
yet whether the additional costs involved in solar improvements should
be excluded from property valuation in determining the maximum
amount they will lend. Only time will tell. In the meanwhile, be aware
of the lender's concern when applying for the mortgage on your new
home. (We refer to the mortgage on your *new* home because in most
cases, loans for solar retrofit of existing homes will be of the home
improvement type. We will review lender's concerns on the home
improvement loans later in this chapter.)

The second major area of uncertainty the lender must evaluate
relates to the technical performance of the system you propose for your
new home. When your solar loan request reaches the lender, the un-

certainties regarding mechanical reliability, energy output, and other technical aspects of the system will have a direct bearing on the value of your property as security for the mortgage. Perhaps not in your mind, or in that of your designer, architect, or mechanical contractor, but financial institutions generally lack the skills (and often the motivation) to examine the technical details of mechanical systems. So be aware, and prepared.

The third major concern which the lender must evaluate is the savings in energy costs to which your system can contribute in future years. Since this will also influence his judgment of your ability to meet future financial obligations, we'll examine the subject a little more closely.

How can savings in energy costs be evaluated? The most direct method is to first establish what your energy costs are before your system is installed; and then what they are after the installation. In theory, the reduction in cost that you enjoy is the result of your new solar heating system. Unfortunately, it's not that simple in real life. That very first step of determining your energy costs before the system installation can be rather bothersome.

If you are one who has kept careful records of your home's fuel usage and cost versus average temperatures over the past five years, the task won't be bad. But for most of us, who just groan and pay our fuel bill when it arrives each month, the task can be formidable. The best source for this information is generally your fuel supplier. Most often, the company or utility firm that supplies your fuel needs will also have accurate records of your home's fuel consumption. When you contact them, ask for at least five to ten years of fuel-use history. This will permit you to *average*, smoothing out the peaks and valleys of particularly cold or mild winters. After arriving at the *average* fuel consumption figure, you're halfway toward knowing the savings (in which your lender is interested).

Next you have to *estimate* that portion of your home's heating needs which your new system can provide. Until you have actual operating experience with your solar system over a period of years, there is no choice but to estimate. Let's assume that the design specifications indicate the system will provide 75% of your heating needs during a "typical" winter season in your geographic location.

At this point, you or your lender can easily calculate the annual savings resulting from the proposed system. Knowing your average fuel usage before the system installation and assuming 75% of that will be saved in each succeeding year makes it easy. Using the most recent fuel price charged, find your savings for the first full year of solar system use, and then increase it by 5% a year (for a conservative fuel price escalation). Here's an example:

> We'll assume your home has an oil-burning furnace, and after checking with your fuel supplier you find that your usage varied between 850 and 1,150 gallons of oil per year, over the past five years, for an average of 1,000 gallons per year. Further assuming that your proposed solar system is designed to provide 75% of your heat, we can conclude that after the installation you will need only 250 gallons of fuel per average year (to maintain the same comfort level). If fuel oil is now selling at 50¢/gallon, your savings of 750 gallons/year is equal to $375 for the first year of solar system operation. By escalating the $375 5% a year, the savings increase from $375 to $394, $413, $434, and to $456 by the end of the fifth year. At that point your total savings exceed $2,000.

This type of information is impressive to a lender who must evaluate your ability to meet future financial obligations. Although the information is fairly accurate, it is not precise. There are many factors which could alter the outcome of your annual savings calculation. To give you some idea, we've listed just a few possibilities.

- The type of fuel used. An all-electric home would be difficult to calculate since the heating system consumes the same basic fuel as the TV, vacuum cleaner, and porch light. How much was used for heating?
- The number of people living in the home.
- The age of persons in the home. Very young or very old people generally need a warmer home during the cold winter months.
- Added insulation will start a "new," reduced, average fuel-use history for any building.

The chart here lists a total of fourteen lender concerns in making loans for solar-heated homes. In addition to the three major points just discussed, there are several lesser ones. Review them, and do your homework before approaching your lending officer for your mortgage loan. Remember, you need be concerned only if you are attempting to secure most of the full cost of your new home and solar installation. If you can afford a reasonable down payment, and the amount you wish to borrow is easily secured by the value of the property, you should have no problem.

ISSUES OF CONCERN TO LENDERS IN MAKING LOANS
Percentage of Lenders Identifying Selected Aspects of Solar Energy Heating Systems as Primary or Substantial Concerns in Future Lending Decisions

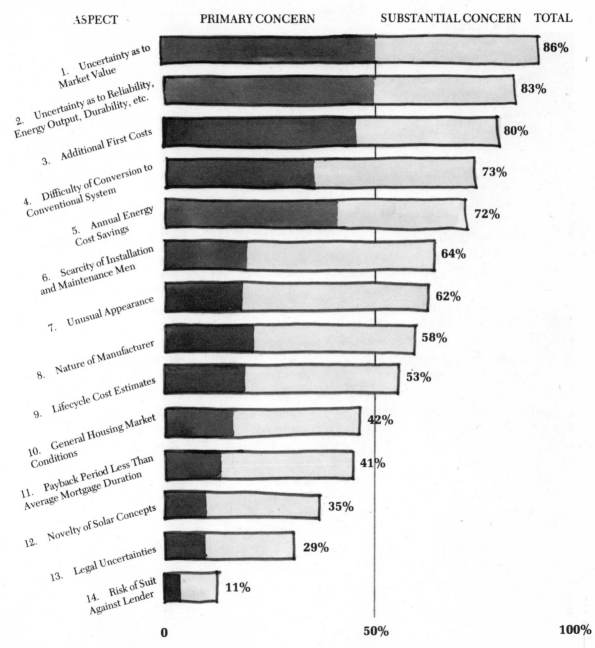

ASPECT   PRIMARY CONCERN   SUBSTANTIAL CONCERN   TOTAL

1. Uncertainty as to Market Value — 86%
2. Uncertainty as to Reliability, Energy Output, Durability, etc. — 83%
3. Additional First Costs — 80%
4. Difficulty of Conversion to Conventional System — 73%
5. Annual Energy Cost Savings — 72%
6. Scarcity of Installation and Maintenance Men — 64%
7. Unusual Appearance — 62%
8. Nature of Manufacturer — 58%
9. Lifecycle Cost Estimates — 53%
10. General Housing Market Conditions — 42%
11. Payback Period Less Than Average Mortgage Duration — 41%
12. Novelty of Solar Concepts — 35%
13. Legal Uncertainties — 29%
14. Risk of Suit Against Lender — 11%

0   50%   100%

Source: The survey conducted in early 1976 by Regional and Urban Planning Implementation, Inc., with support from the National Science Foundations.

229

Our discussion of lender's concerns has examined the problems of securing a loan for the person who is purchasing or building a new home. We've reviewed some of the questions the lender is sure to ask. But what about the borrower who wants to add a solar retrofit to his present home?

Here, the lender's concern is limited almost exclusively to the borrower's credit worthiness (the amount of additional indebtedness he or she can reasonably be expected to support). A review of the typical home improvement loan application earlier in the chapter will highlight, once again, the information you are requested to supply in order that the lender can evaluate your credit standing.

FINANCING'S IMPACT ON THE FUTURE USE OF SOLAR HEATING

In spite of all the foregoing discussions, it's all likely to hinge on what the builders and developers, as opposed to the individual home buyers, do. If more energy-conscious buyers look for fuel-saving features—and are willing to pay the higher first costs involved—builders, and the lending institutions that support them, will respond accordingly. The builders and developers of single-family housing may well be the "foot-in-the-door" that opens the mortgage market to solar homes.

Initially, the mortgage holders will be more likely to finance a developer, rather than a home buyer, because the developer will be "putting his money where his mouth is." The developer will have applied *his* professional expertise to evaluating both the technical performance of the solar system he is using, and the marketability of the resulting home. And the lender knows this. Probably the lender will restrict the builder to only one or two "speculative" solar homes in his first development, until the results prove the value is there and the homes are salable. However, as the economics of solar use become accepted, the financing of systems in multi-family housing may spread even faster than in the arena of single-family homes.

A recent article which appeared in the *Christian Science Monitor* permits us to close this chapter on a note of optimism. See if you don't agree.

financial

Two banks jump into financing solar buildings in California

By Lewis Brigham
Special to The Christian Science Monitor

San Francisco

Two major California banks have shoved traditional real estate loan caution aside and appear to have become the nation's first commercial banks to underwrite solar energy financing. The loans are for a 33-unit housing tract just west of Palm Springs, and an industrial park in Santa Clara.

Not long ago, two Boston savings banks – Charlestown and Suffolk Franklin Savings – included solar energy in their joint financing of a cooperative housing venture on 85 acres in North Easton, Massachusetts. But these are real estate-oriented savings institutions – not commercial banks.

The two innovators here are Bank of America and Wells Fargo. Wells Fargo supplied $2 million in construction financing for the Oakmead industrial park in Santa Clara, which includes the nation's first two industrial structures designed from the ground up to depend entirely on solar energy.

And for long-term financing of the structures, Wells Fargo Mortgage also arranged a $2.9 million investment with Philadelphia Savings Fund Society.

Lacking figures on costs of solar energy equipment in relation to savings they will generate, Wells Fargo extended financing "as if the project were to be conventionally heated," said Chris K. Williams, the bank's officer handling the loan. But he added that "as the cost of energy goes up, the value of the finished product may far exceed the present loan commitment." The two structures will be completed in August.

The Bank of America venture into single-family tract housing represents an even more innovative step for a commercial bank. It is believed the bank became the first U.S. bank to finance solar homes on a tract basis, and also the first to extend long-term loans on such homes with a commitment to provide mortgage money.

The development is located in Hemet, California, a middle-income retirement community 40 miles west of Palm Springs. Seventeen medium-priced family homes have already been built and sold, and the remaining 16 will be completed in August.

The concept originated with Warren Buckmaster, a diamond company executive in New York City up to 1974 when he took $110,000 in savings and moved to Hemet, California. He spent the first 18 months studying, sounding out suppliers, and talking to experts in home design, construction, and solar energy.

One of the construction experts was W. F. MacDonald, a San Diego-based engineer-builder. Together they formed Blue Skies Radiant Homes. Their goal was to design a house like a refrigerator, literally enveloped by insulation to retain its atmosphere regardless of the temperature outside.

They also decided to incorporate a simple solar system based on the hydronic forced-air principle where water is pumped to the roof and heated to about 160 degrees, then pumped down into a 1,000-gallon insulated cement storage tank. Water intended for hot water or heating comes into the house through copper coils in the storage tank.

The Blue Skies units are Spanish-style stucco structures with red tile roofs, two-car garages, and parapets on the roofs which hide the solar panels from view. The first 17 homes were sold in two weeks, and were priced at $37,900 for two bedrooms, $45,900 for three, and $46,900 for four. The prices were comparable to nonsolar units in Hemet and an astonishing bargain by Los Angeles standards.

But long before the first foundation was laid, the developers spent eight months trying to find financial backing for the venture. Before Bank of America's entry into the picture, Mr. Buckmaster said "I went to perhaps 40 other southern California banks and lending institutions without success."

The senior vice-president of one notable Los Angeles firm wrote Mr. Buckmaster that he believes in radiant heating, but added "our conventional financing sources feel you are ahead of your time."

John M. Nead, vice-president for Bank of America's real estate loan department in Los Angeles, didn't see it that way. "I decided we should take the chance. I believe that the use of solar energy on a large scale is coming and here was a chance to get involved," he noted.

The bank's funding amounted to $540,000 for the first 17 homes and $580,000 for the second group with a commitment of $550,000 for mortgages.

APPENDIX/CHAPTER 9

BIBLIOGRAPHY

Home Mortgage Lending and Solar Energy. Regional and Urban Planning Implementation, Inc., Cambridge, Mass. David Barrett, Peter Epstein, Charles M. Haas, February 1977.

Money and Banking.   Steiner, Shapiro, Solomon, New York: Holt, Rinehart and Winston, Inc., 1958, 4th edition.

CONGRATULATIONS!

You have just read the whole book. Now you'll no longer have us at your elbows, whispering advice into your ears each step of the way. You're alone again, out in the cold, cruel world of solar sharks and energy wolves. But with the protective shield we've given you, you can swim and walk among them with confidence. With luck, your solar adventure can be a delight, and your can help an entire nation end its frenzy of fuel burning.

Before we leave you, however, we'd like to call your attention to the pages that follow. They include the general appendices and start with a checklist; a memory-jogger to remind you of the highlights of the previous chapters. Use it as you plan—or actually begin your solar project. If it prevents you from making one needless error it will have served its purpose well. Between the information offered in this general Appendix and the information to which they lead, the whole new world of appropriate technology is at your fingertips.

Good luck!

—Irwin Spetgang and Malcolm Wells

GENERAL APPENDIX

Reader's Checklist

Chapter 2—Basic Training

Item   Comment

1. Do you understand the difference between an active and a passive solar system?
2. Are there any questions unanswered in your mind regarding the basic principals of solar heating?
3. If so, check your local library for any of the text material in the bibliography at the end of the chapter.
4. Have you located your region on the U.S. map found on page 12? The chart on page 13 indicates the percentage of fuel you can reasonably expect to save by using solar heating in the region where you live.
5. Do you know the collector area and heat-storage tank size recommended for your home region? (See page 21 for this information.)
6. The chart on page 29 provides a range of dollar costs for solar heating system installations. Have you checked these costs for your region?

Chapter 3—Insulation

1. Check for *all* air leaks in your home.
2. Have you checked the condition of your attic insulation?
3. How about the basement or crawl space?
4. Talk with a representative of your local building supply business. He can provide you with valuable "how to do your own insulation" information.
5. Be sure to dress properly before attempting insulation work.
6. Create your own check list concerning the tools, equipment, materials, and protective clothing needed for insulating your home.
7. Do you understand the need for vapor barriers, and where they should be used?
8. Have you planned your insulation work carefully?
9. Get competitive quotes on the materials needed.
10. Examine all the windows and doors in your home. What can you do to improve them?
11. Consider the insulating effects of shutters, shades, drapes, storm windows, etc.

Item	**Comment**

12. Can air-lock vestibules be added conveniently?
13. Have you checked the list of government publications available on insulation, and have you ordered any?
14. Survey and list all hot and cold water pipes that are accessible, and see if they can be insulated.
15. Is your domestic hot water storage tank insulated on the outside (in addition to the manufacturer's inner insulation)?

Chapter 4—Your Home's Solar Potential

1. Do you know the age of your home? Check your property settlement papers for that data. While you're at it, check for the original plot plan which shows the north direction arrow.
2. Have you ever counted the number of windows in your home? Refer to page 83 and refresh your memory by rereading the questions asked about the number and types of doors and windows in your home.
3. Is the axis of your home oriented close to an east-west line?
4. The deviation of *magnetic north* from *true north* can be found in the reference area of your local library. Ask your librarian for help if you need it.
5. Is there a logical place to mount solar collectors on your home, in the quadrant that gets maximum exposure to sunlight?
6. Do you know the geographic latitude of your property?
7. How about the slope of your roof? Do you know how great it is?
8. How many square feet of heated area are there in your home?
9. Is your present heat distribution system suitable for conversion to solar heat distribution as well?
10. Is your domestic hot water heating system separate from your home's space heating system?
11. Have you reviewed all possible installation restrictions, both indoors and out?
12. Where would you locate your energy storage tank (or rock-bin)?

Chapter 5—Your Conservation Habits

This chapter is a total check list in itself. So if you haven't checked the answers to all of the questions asked, go back and do it now.

Chapter 6—Our Solar Poll

Item   **Comment**

1. The only item we'd suggest with regard to this chapter is that you visit the solar home nearest you. Plan ahead, get permission for your visit, and then ask only a few meaningful questions after you get there. If that isn't possible, see if any interviews of solar home owners can be found in your library.

Chapter 7—Contracts and Contractors

1. Have you requested *recent* references from each competing contractor?
2. And have you checked these references?
3. Get at least three (3) competing bids.
4. Are your bids "firm prices" as opposed to "estimates?"
5. Don't forget the need for a building permit. Who will obtain it, you or your contractor?
6. Will your contractor notify the building inspection department when such notification (work ready for inspection) is required?
7. Will you need help in specifying (writing the specifications) the work you want?
8. Review the sample contract forms used as illustrations in this chapter.
9. Did you compare the terms of the AIA contract starting on page 166, and those of a contractor, page 174?
10. Have you set up and maintained a log or diary of all communications with your contractor? You'll be glad you did.

Chapter 8—Legal

1. Is there, adjacent to your property, undeveloped land on which future construction might block your sunshine?
2. Are you familiar with your community's zoning regulations?
3. And with the building codes?
4. Are building permits required in your community?
5. Must you notify the inspectors as you complete portions of the installation?
6. Is an occupancy permit needed before you can use your new system?
7. Are you familiar with your community's zoning variance process?

Item   Comment

8. Does your state have tax incentives to encourage the use of
   solar heating?
9. Will a solar installation increase the assessed value of—and
   the taxes on—your property?

Chapter 9—Financial

1. Have you an established credit rating at a local bank?
2. Have you "shopped" for the best rate available from your local
   lending institutions?
3. Review the list of lenders' concerns. Then be sure you have
   answers for your lender's questions—before he asks them.
4. Have you kept records of the fuel used by your conventional
   heating system?
5. Are your "pay-back" calculations completed?

Manufacturers of Solar Collectors*†
January-June 1976

This key explains the types of solar collectors in which each of the listed firms specializes.

L,L = Low temperature, liquid
M,A = Medium temperature, air
M,L = Medium temperature, liquid
S,L = Special Collector, liquid

S,L   AAI
      P.O. Box 6767
      Baltimore, MD 21204
      301/666-1400

M,L   Advance Cooler Manufacturing
      Corp.
      Route 146, Bradford Industrial
      Park
      Clifton Park, NY 12065
      518/371-2140

S,L   Albuquerque Western
      Industries, Inc.
      612 Commanche, N.E.
      Albuquerque, NM 87107
      505/344-7224

M,L   Alten Associates, Inc.
      3080 Olcott Street—Suite 200D
      Santa Clara, CA 95051
      408/247-6967

M,A   Amcon, Inc.
      211 W Willow Street
      Carbondale, IL 62901
      618/457-3022

*Excerpted from NTIS document #PB 258 865, prepared by Richard Stoll of the Federal
Energy Administration, Sept. 1976.

†For a listing of manufacturers selling solar hot water equipment, see Appendix/Chapter 5.

M,L American Helio Thermal Corp.
3515 South Tamarac Drive
Denver, CO 80237
303/773-6085

L,L American Solar Heat Corporation
M,L 4210 Peters Road
Fort Lauderdale, FL 33317
305/974-2500

M,L American Solar King Corp.
6801 New McGregor Highway
Waco, TX 76710
817/776-3860

M,L American Solar Power, Inc.
5018 West Grace Street
Tampa, FL 33607
813/879-4943

M,L Ametek, Inc., Power Systems
Group
1 Spring Avenue
Hatfield, PA 19440
215/822-2971

L,L Atlas Vinyl Products
7002 Beaver Dam Road
Levittown, PA 19057
215/946-3620

M,L Bay Area Solar Collectors
3068 Scott Blvd.
Santa Clara, CA 95050
408/985-2272

M,L Beutels Solar Heating Company
7161 N.W. 74th Street
Miami, FL 33166
305/885-0122

M,L Bright Industries Sun Prod. Inc.
1900 N.W. 1st Court
Boca Raton, FL 33432
305/391-4686

L,L Burke Industries, Inc.
2250 South 10th Street
San Jose, CA 95112
408/297-3500

M,A C & C Solar Thermics
Box 144
Smithsburg, MD 21783
301/631-1361

M,L Calmac Manufacturing Corp.
P.O. Box 710E
Englewood, NJ 07631
201/569-0420

M,L Capital Solar Heating, Inc.
376 N.W. 25th Street
Miami, FL 33127
305/576-2380

M,L Carolina Solar Equipment Co.
P.O. Box 2068
Salisbury, NC 28144
704/637-1243

M,L Chamberlain Mfg. Corp.
845 Larch Avenue
Elmhurst, IL 60126
312/279-3600

M,A Champion Home Builders
5573 E. North Street
Dreyden, MI 48428
313/796-2211

M,L Chemical Processors, Inc.
P.O. Box 10636
St. Petersburg, FL 33733
813/822-3689

M,L Columbia Solar Energy Div.
L,L 55 High Street
Holbrook, MA 02343
617/767-0513

M,L Consumer Energy Corporation
4234 S.W. 75th Avenue

Miami, FL 33155
305/266-0124

M,A Contemporary Systems, Inc.
68 Charlonne Street
Jaffrey, NH 03452
603/532-7972

M,A Crimsco, Inc.
5001 East 59th Street
Kansas City, MO 64130
816/333-2100

M,L CSI Solar Systems Division
12400 49th Street
Clearwater, FL 33520
813/577-4228

M,L D.W. Browning Contracting Co.
475 Carswell Avenue
Holly Hill, FL 32017
904/252-1528

M,L Daystar Corporation
41 Second Avenue
Burlington, MA 01803
617/272-8460

M,L Dayton Sanitary Engineering Co.
P.O. Box 1444
Dayton, OH 45414
513/278-6551

S,L Del-Jacobs Engineering
837 S. Fair Oaks Avenue
Pasadena, CA 91105
213/681-4561

M,L Dick Mills/Airtron Inc.
15286 U.S. Highway S.
Clearwater, FL 33516
813/531-3581

S,L Diversified Natural Resources
8025 East Roosevelt
Scottsdale, AZ 85257
602/945-2330

M,L E&K Service Company
16824 74th Avenue N.E.
Bothell, WA 98011
206/486-6660

M,A El Camino Solar Systems
5330 Debbie Lane
Santa Barbara, CA 93111
805/967-6527

M,L Energy Converters, Inc.
2501 N. Orchard Knob Avenue
Chattanooga, TN 37406
615/624-2608

M,L Energy Dynamics Corporation
327 West Vermijo Road
Colorado Springs, CO 80903
303/475-0332

M,L Energy Systems, Inc.
634 Crest Drive
El Cajon, CA 92021
714/447-1000

M,L Enviropane, Inc.
348 N. Marshall Street
Lancaster, PA 17602
717/299-3737

L,L FAFCO
138 Jefferson Drive
Menlo Park, CA 94025
415/321-6311

S,L Falbel Energy Systems Corp.
472 Westover Road
Stamford, CT 06902
203/357-0626

M,L Flagala Corporation
9700 W. Highway 98
Panama City, FL 32401
904/234-6559

M,L Florida Solar Power, Inc.
1327 South Monroe Street

P.O. Box 5846
Tallahassee, FL 32301
904/224-8270

M,L General Electric Company
P.O. Box #8661/Bldg. #7
Philadelphia, PA 19101
215/962-4785

M,L General Energy Devices
1753 Ensley
Clearwater, FL 33516
813/586-1146

M,L Grumman Aerospace Corp.
Energy Programs, Plant 25
Bethpage, NY 11714
516/575-6205

M,L Gulf Thermal Corporation
2215 Industrial Blvd.
P.O. Box 13124 Airgate Branch
Sarasota, FL 33580
813/355-9783

M,L Halstead Industries
P.O. Box 1110
Scottsboro, AL 35768
205/259-1212

M,L Hansberger Refrigeration &
Electric Co.
2450 8th Street
Yuma, AZ 85364
602/783-3331

M,L Helio Associates
P.O. Box 17960
Tucson, AZ 85731
602/792-2800

M,L Heliotherm, Inc.
Lenni, PA 19052
215/459-9030

S,L Hexcel Corporation
11711 Dublin Blvd.

Dublin, CA 94566
415/828-4200

M,A Howard Bell Enterprises, Inc.
P.O. Box 413
Valley City, OH 44280
216/483-3249

M,L Hughes Supply, Inc.
P.O. Box 2273
Orlando, FL 32802
305/841-4710

M,L Interactive Resources, Inc.
39 Washington Avenue
Point Richmond, CA 94801
415/236-7435

M,L International Environment Corp.
129 Halstead Avenue
Mamaroneck, NY 10543
914/698-8130

M,L International Solar Heating
Corporation
P.O. Box 55
Front Royal, VA 22630
703/635-3337

M,A J.G. Johnston Company
33458 Angeles Forest Hwy.
Palmdale, CA 93550
805/947-3791

M,L Kennecott Copper Corporation
128 Spring Street
Lexington, MA 02173
617/862-8268

M,L KTA Corporation
12300 Washington Avenue
Rockville, MD 20852
301/468-2066

M,L Largo Solar Systems, Inc.
2525 Key Largo Lane
Fort Lauderdale, FL 33312
305/583-8090

M,L   Lennox Industries, Inc.
      200 S. 12th Avenue
      Marshalltown, IA 50158
      515/754-4011

M,L   Libby Owens Ford
      Technology Center
      1701 East Broadway
      Toledo, OH 43605
      419/247-4355

M,L   Martin Marietta Corporation
      P.O. Box 179, Mail Stop 0202
      Denver, CO 80201
      303/979-7000

S,L   Mechanical Mirror Works
      661 Edgecombe Avenue
      New York, NY 10032
      212/795-2100

M,L   National Solar Supply
      2331 Adams Drive N.W.
      Atlanta, GA 30318
      404/352-3478

M,L   Natural Energy Systems
      Marketing Arms Division
      1632 Pioneer Way
      El Cajon, CA 92020
      714/440-6411

S,L   Northrup, Inc.
M,L   302 Nichols Drive
      Hutchins, TX 75141
      214/225-4291

S,L   Owens Illinois, Inc.
      P.O. Box 1035
      Toledo, OH 43666
      419/242-6543

L,L   P.R. Distributors
      1232 Zacchini Avenue
      Sarasota, FL 33577
      813/958-5660

M,L   Paul Meuller Company
      P.O. Box 828
      Springfield, MO 65801
      417/865-2831

M,L   Piper Hydro Corporation
      2895 East La Palma
      Anaheim, CA 92806
      714/630-4040

M,L   Powell Brothers, Inc.
      5903 Firestone Blvd.
      South Gate, CA 90280
      213/869-3307

M,L   PPG Industries
      One Gateway Center
      Pittsburgh, PA 15222
      412/434-3552

M,L   RAYPAK, Inc.
L,L   3111 Agoura Road
      Westlake Village, CA 91361
      213/889-1500

M,L   Research Engineering
      Manufacturing Co.
      5285 North Redrock Drive
      Phoenix, AZ 85018
      602/959-0761

M,L   Revere Copper & Brass, Inc.
      P.O. Box 151
      Rome, NY 13440
      315/338-2022

M,L   Reynolds Metal Company
      6601 West Broad Street
      Richmond, VA 23261
      804/281-3026

M,L   Rocky Mountain Air
      Conditioning
      5010 Cook Street
      Denver, CO 80216
      303/825-0203

M,L S.W. Ener-Tech, Inc.
3030 S. Valley View Blvd.
Las Vegas, NV 89102
702/873-1975

M,L Scientific-Atlanta, Inc.
3845 Pleasantdale Road
Atlanta, GA 30340
404/449-2000

M,L SEMCO
1091 S.W. 1st. Way
Deerfield Beach, FL 33441
305/427-0040

S,L Sheldahl Company
P.O. Box 170
Northfield, MN 55057
507/645-5631

L,L Sigma Energy Products
M,L 1405 B San Mateo N.E.
Alburquerque, NM 87110
505/262-0516

M,L Solamatic
220 West Branden Blvd.
Branden, FL 33511
813/689-1182

M,A Solar Aire
M,L 1611 W. 9th Street
White Bear Lake, MN 55110
612/429-3844

M,L Solar Central
7213 Ridge Road
Mechanicsburg, OH 43044
513/828-1350

M,L Solar Corporation of America
19 Winchester Street
Warrenton, VA 22186
703/347-0550

M,L Solar Development, Inc.
M,A 4180 West Roads Drive

West Palm Beach, FL 33407
305/842-8935

M,L Solar Dynamics, Inc.
4527 E. 11th Avenue
Hialeah, FL 33013
305/688-4393

M,L Solar Energy Components, Inc.
1605 North Cocoa Blvd.
Cocoa, FL 32922
305/632-2880

M,L Solar Energy Contractors
3156 Leon Road
P.O. Box 17094
Jacksonville, FL 32216
904/641-5611

M,A Solar Energy Products Company
121 Miller Road
Avon Lake, OH 44012
216/933-5000

M,L Solar Energy Research Corp.
1228 15th Street
Denver, CO 80202
303/573-5499

M,L Solar Energy Resources Corp.
10639 SW 185 Terrace
Miami, FL 33157
305/233-0711

M,L Solar Energy Systems
1243 South Florida Avenue
Rockledge, FL 32955
305/632-6251

L,L Solar Energy Systems, Inc.
M,L 2492 Banyan Drive
Los Angeles, CA 90049
213/472-6508

M,L Solar Energy Systems, Inc.
9016 Collins Avenue
Pennsauken, NJ 08110
609/665-6800

M,L Solar Engineering, Inc.
P.O. Box 1358
Boca Raton, FL 33432
305/368-2456

L,L Solar Enterprises
9803 E. Rush Street
El Monte, CA 91733
213/444-2551

M,L Solar Heat Corporation
1252 French Avenue
Lakewood, OH 44107
216/228-2993

M,L Solar Heating & Air Conditioning
Systems
13584 49th Street North
Clearwater, FL 33520
813/577-3961

M,L Solar Heating Systems Corp.
151 John Downey Drive
New Britain, 06051
203/224-2164

M,L Solar Industries of Florida
P.O. Box 9013
3231 Trot River Blvd.
Jacksonville, FL 32216
904/768-4323

M,L Solar One Ltd.
709 Birdneck Road
Virginia Beach, VA 23451
804/422-3262

S,L Solar Physics Corporation
M,L 1350 Hill Street, Suite A
El Cajon, CA 92020
714/440-1625

L,L Solar Pool Heaters of SW Florida
901 S.E. 13th Place
Cape Coral, FL 33904
813/542-1500

M,L Solar Products, Sun-Tank, Inc.
614 N.W. 62nd Street

Miami, FL 33150
305/756-7609

M,L Solar Research/Refrigeration
Research
525 N Fifth Street
Brighten, MI 48116
313/227-1151

M,A Solar Shelter
P.O. Box 36
Reading, PA 19603
215/488-7624

M,L Solar Systems
54 Ervin Street
Belmont, NC 28012
704/825-8416

M,L Solar Systems by Sundance
10021 SW 38th Terrace
Miami, FL 33165
305/221-4611

S,L Solar Systems, Inc.
507 W. Elm Street
Tyler, TX 75701
214/592-5343

M,L Solar Utilities
11404 Sorrento Valley Road
San Diego, CA 92121
714/452-8822

M,L Solaray, Inc.
M,A 324 S. Kidd Street
Whitewater, WI 53190
414/473-2525

M,L The Solaray Corporation
2414 Makiki Heights Drive
Honolulu, HI 96822
808/533-6464

M,L Solarcoa
4647 Long Beach Blvd. N.
Long Beach, CA 90805
213/426-7655

M,L Solargizer Corporation
220 Mulberry Street
Stillwater, MN 55082
612/739-0117

M,A Solaron Corporation
4850 Olive Street
Commerce City, CO 80022
303/289-5971

M,L Solarway
P.O. Box 217
Redwood Valley, CA 95470
707/485-7616

M,L Southeastern Solar Systems
4705 J Bakers Ferry Road
P.O. Box 44066
Atlanta, GA 30336
404/691-1864

M,L Southern Lighting Mfg. Co.
501 Elwell Avenue
Orlando, FL 32803
305/894-8851

M,L Standard Electric Co.
P.O. Box 631
Rocky Mount, NC 27801
919/442-1155

M,L State Industries, Inc.
Cumberland Street
Ashland City, TN 37015
615/792-4371

M,L Sun Century Systems
P.O. Box 2036
Florence, AL 35630
205/764-0795

L,L The Sundu Company
3319 Keys Lane
Anaheim, CA 92804
714/828-2873

M,L Sun Earth, Inc.
Progress Drive

Montgomeryville, PA 18736
215/699-7892

M,A Sun Stone
P.O. Box 941
Sheboygan, WI 53081
414/452-8194

M,L Sun Systems of America, Inc.
P.O. Box 10336
Jacksonville, FL 32207
904/389-0493

M,L Sun Systems, Inc.
P.O. Box 347
Milton, MA 02186
617/268-8178

L,L Sunburst Solar Energy, Inc.
M,L P.O. Box 816
Menlo Park, CA 94025
415/595-0440

S,L Sunpower Systems Corporation
2123 S. Priest Road, Suite 216
Tempe, AZ 85282
602/968-6387

M,L SunSav, Inc.
250 Canal Street
Lawrence, MA 01840
617/686-8040

M,L Sunseeker Systems, Inc.
100 W Kennedy Blvd.
Tampa, FL 33602
813/223-1787

M,L Sunshine Utility Company
1466 Pioneer Way, Suite 3
El Cajon, CA 92020
714/440-3151

M,A Sunwall, Inc.
P.O. Box 9723
Pittsburgh, PA 15229
412/364-5349

M,L Sunworks Division/Enthone Co.
M,A P.O. Box 1004
     New Haven, CT 06508
     203/934-6301

M,L Temp-O-Matic Cooling Co.
     87 Luguer Street
     Brooklyn, NY 11231
     212/624-5600

M,L Tranter
     735 East Hazel Street
     Lansing, MI 48909
     517/372-8410

L,L U.S. Solar Pillow
     P.O. Box 987
     416 East Oak
     Tucumcari, NM 88401
     505/461-2608

M,L Unit Electric Control, Inc.
     130 Atlantic Drive
     Maitland, FL 32751
     305/831-1900

M,L Universal Solar Energy Company
     1802 Madrid Avenue
     Lake Worth, FL 33461
     305/586-6020

M,L W.R. Robbins & Sons
     1401 N.W. 20th Street
     Miami, FL 33142
     305/325-0880

M,L Wallace Company
     831 Dorsey Street
     Gainsville, GA 30501
     404/534-5971

M,L Western Energy, Inc.
     454 Forest Avenue
     Palo Alto, CA 94302
     415/327-3371

M,L Wilcon Corporation
     3310 S.W. Seventh
     Ocala, FL 32670
     904/732-2550

M,L Wilcox Manufacturing Corp.
     13375 U.S. 19 North
     P.O. Box 455
     Pinellas Park, FL 33565
     813/531-7741

S,L Wilson Solar Kinetics Corp.
     P.O. Box 17308
     West Hartford, CT 06117
     203/233-7884

M,L Ying Manufacturing Corp.
     1940 West 144th Street
     Gardena, CA 90249
     213/327-8399

M,L Youngblood Company, Inc.
     1085 NW 36th Street
     Miami, FL 33127
     305/635-2501

Solar Kit Manufacturers

A to Z Solar Products
200 East 26th Street
Minneapolis, MN 55404
612/870-1323
Model 22 Solar Domestic Water Heating
    System and Many Other Devices.

Diy-Sol, Inc.
P.O. Box 614

Marlboro, MA 01752
Do-It-Yourself System

Ecotope Group
Box 618
Snohomish, WA 98290
Domestic Hot Water System

J & B Sales Co.
3441 North 29th Avenue

Phoenix, AZ 85017
Domestic Hot Water System and Collectors

Solar Corporation of America
19 Winchester Street
Warrenton, VA 22186
703/347-0550
Swimming Pool Heater Kit

Solar Energy Digest
P.O. Box 17776
San Diego, CA 92117
714/277-2980
Solarsan Solar Water and Space Heaters

Solargenics, Inc.
9713 Lurline Avenue
Chatsworth, CA 91311
213/998-0806
Domestic Hot Water System

Solar Research
Refrigeration Research
525 North Fifth Street
Brighton, MI 48116
313/227-1151
SR 501 Solar Domestic Water
  Heating System

Solar Directories*

ARIZONA SOLAR ENERGY DIRECTORY (1976)
(free to Arizona residents; $1.00 outside Arizona)

> Solar Energy Research Commission
> 1700 West Washington St.
> Room 502
> Phoenix, AZ 85007
> 602/271-3682

DIRECTORY OF SOLAR ENERGY EQUIPMENT MANUFACTURERS AND
SOLAR ENERGY ARCHITECTS AND ENGINEERS (1976, free)

> Department of Business and Economic Development, Division of Energy
> 222 South College Street
> Springfield, IL 62706
> 217/782-7500

DIRECTORY OF THE SOLAR INDUSTRY (1976, $7.50)

> Solar Data
> 13 Evergreen Road
> Hampton, NH 03842
> 602/926-8082

*Compiled by National Solar Heating and Cooling Information Center, P.O. Box 1607, Rock-
ville, MD 20850.

FLORIDA SOLAR ENERGY EQUIPMENT AND SERVICE (1976, free)

Florida Solar Energy Center
300 State Road 401
Cape Canaveral, FL 32920
305/783-0300

SOLAR AGE CATALOG (1977, $8.50)

Solar Age
P.O. Box 305
Dover, NJ 07801

SOLAR COLLECTOR MANUFACTURING ACTIVITY (1976, free)

National Energy Information Center
Federal Energy Administration
Federal Building, Room 4358
12 & Pennsylvania Ave., N.W.
Washington, DC 20461
202/961-8685

SOLAR DIRECTORY (1975, $20.00)

Carolyn Pesko
Ann Arbor Science Publishers
Box 1425 Ann Arbor, MI 48106
313/761-5010
N.B.: A new version will be published, July 1977 under the title
  SOLAR ENERGY AND RESEARCH DIRECTORY

SOLAR DIRECTORY-VOLUME 1 (1977, $20.00)

Home Free
4924 Greenville Ave.
Dallas, TX 95206
214/368-8850

SOLAR ENERGY DIRECTORY (1976, $7.50)

Centerline Corporation
401 S. 36th Street
Phoenix, AZ 85034
602/267-0014

SOLAR ENERGY INDEX (1977, $8.00)

Solar Energy Industries Association
1001 Connecticut Ave. N.W.

Washington, DC 20036
202/293-2981

SOLAR SOURCE BOOK (1977, $12.00 prepaid)

SEINAM (Solar Energy Institute of America)
Dept. C
P.O. Box 9352
Washington, DC 20005
202/853-2335

SPECTRUM (1975, $2.00)

Alternative Sources of Energy
Route 2, Box 90A
Milaca, MN 56353
612/983-6892

WESTERN REGIONAL SOLAR ENERGY DIRECTORY (1976, $2.35)

Southern California Solar Energy Association
Chapter International Solar Energy Society (I.S.E.S.)
202 C Street–11B
San Diego, CA 92101
714/232-3914

1977 SUN CATALOGUE ($2.00)

Solar Usage Now
Box 306
Bascom, OH 44809
419/937-2226

Federal Publications on Solar Energy*

Buying Solar. 1976. 80 p.

| | FE 1.2:So4/3 | 041/018-00120-4 | $ 1.85 |

Central Receiver Solar Thermal Power System, Phase 1, 10-MW Electric Pilot Plant. 1976. 28 p. il.

| | ER 1.2:So 4/7/Phase 1 | 060-000-00009-3 | .90 |

Converting Solar Energy Into Electricity: A Major Breakthrough? Hearing, 94th Congress, 2d Session, June 11, 1976. 1976, reprinted 1977. 38 p.

| | Y 4.G 74/7:So 4 | 052-070-03525-1 | .55 |

*Available from the Superintendent of Documents, United States Government Printing Office, Washington, DC 20402.

Definition report, National Solar Energy Research, Development and
    Demonstration Program. 1976. 88 p.

          ER 1.11:ERDA-49   052-010-00473-5 $ 2.00

Demand Analysis, Solar Heating and Cooling of Buildings. This report
    focuses on two separate subjects relating to the use and accep-
    tance of solar heating and cooling by the general public. Part one
    is a summary of the solar water heater industry in Miami, Florida.
    It describes products and technology of the solar water heater in-
    dustry, economics of marketing solar water heaters, users' cost,
    and opinions of Miami residents on solar water heaters. Part two
    reviews the attitudes of lending institutions and financiers
    throughout the country toward the solar heating and cooling of
    buildings. 1975, reprinted 1976. 169 p. il.

          NS 1.33:So 4/3   038-000-00207-4   2.45

Development of Proposed Standards for Testing Solar Collectors and
    Thermal Storage Devices. 1976. 268 p. il.

          C 13.46:899   003-003-01579-9   3.10

An Economic Analysis of Solar Water and Space Heating.

          In Preparation.   060-000-00038-7   1.40

Energy Research and Development Administrations Pacific Regional
    Solar Heating Handbook. 1976.

          ER 1.8:504   060-000-00024-7   3.25

Energy Research and Development and Small Business, Hearings
    Before the Select Committee on Small Business, Senate, 94th
    Cong., 1st Session, May 13, and 14, 1975, on Solar Energy: How
    Much? How Much from Small Business? How Soon? Why Not
    More? Why Not Sooner?

Pt. 1. 1975. 809 p. il.

          Y 4.Sm 1/2:En 2/pt. 1   052-070-03108-5   6.60

Pt. 1A, Appendixes. 1975. p. 811-1564, il.

          Y 4.Sm 1/2:En 2/pt. 1A   052-070-03107-7   6.20

Pt. 1B, Appendixes. 1975. p. 1565-2623, il.

          Y 4.Sm 1/2:En 2/pt. 1B   052-070-03109-3   9.60

Pt. 1C, Appendixes. 1975. p. 2625-3766, il.

          Y 4.Sm 1/2:En 2/pt. 1C   052-070-03110-7   10.25

Pt. 1D, Appendixes. 1975. p. 3767-4329, il.

          Y 4.Sm 1/2:En 2/pt. 1D   052-070-03111-5   4.85

Pt. 2, Solar Energy Continued The Small Business and Government Roles, October 8, 22, and November 18, 1975. 1976. p. 4330-5460, il. $ 10.00

Pt. 2A Appendixes. 1976. p. 5462-5832, il.
Y 4.Sm 1/2:En 2/pt. 2A   052-070-03700-8   3.40

Energy Research and Technology, Abstracts of NSF/RANN Research Reports, October 1970-December 1974. 1975. 319 p.
NS 1.13/3-2:75-6   038-000-00330-5   3.80

Energy, Solar Heating and Cooling Systems in Buildings, Memorandum of Understanding Formulated at Odeillia, France, Entered Into Force July 1, 1975. 1976. 4p.
S 9.10:8202   044-000-90755-1   .35

Energy Use and Climate: Possible Effects of Using Solar Energy Instead of "Stored Energy." A brief study of the possible effects on worldwide climate of using solar energy instead of "stored" energy (such as fossil and nuclear fuel) to meet future energy needs. 1975, reprinted 1976. 37 p. il.
NS 1.2:En 2/5   038-000-00240-6   1.35

High Energy Phenomena on the Sun, A Symposium Held at Goddard Space Flight Center, Greenbelt, Maryland, September 23-30, 1972. Contains 50 papers which were presented at the symposium. 1973. 641 p. il.
NAS 1.21:342   033-000-00545-8   5.75

Industry Opinions on the Formation of a Solar Energy Research Institute (SERI). Information contained in the study includes background, samples of industry opinion, and results and recommendations for the SERI. 1975. 50 p. il.
ER 1.11:Mtr-7067   052-010-00458-1   1.60

Inexpensive Economical Solar Heating System for Homes. 1976. 56 p. il.
NAS 1.15:X-3294   033-000-00632-2   1.15

Interim Performance Criteria for Solar Heating and Combined Heating/Cooling Systems and Dwellings. 1975. 110 p. il.
C 13.6/2:So 4   003-003-01388-9   1.90

National Program Plan for Research and Development in Solar Heating and Cooling, Interim Report. 1976. 186 p. il.
ER 1.11:ERDA 76-144   060-000-00026-3   2.70

National Program Plan for Solar Heating and Cooling of Buildings, Project Data Summaries:

Volume 1, Commercial and Residential Demonstrations, 1976. 163 p. 81.

ER 1.11:ERDA 76-127	060-000-00012-3	$ 2.35

Volume 2, Demonstration Report. 1977. 68 p.

ER 1.11:ERDA 76-128	060-000-00042-5	1.25

Volume 3, Research and Development, 1976. 128 p.

ER 1.11:ERDA 76-145	060-000-00018-2	1.90

National Program for Solar Heating and Cooling of Buildings, Annual Report.

In Preparation.	060-000-00043-3	1.55

National Program for Solar Heating and Cooling, (Residential and Commercial Applications). Discusses the Federal portion of the program, which is designed to stimulate commercial interest in producing and distributing solar heating and cooling systems. Chapters include working examples of and research in solar heating and cooling, plus the strategy and goals of the Energy Research and Development Administration (ERDA). 1976. 83 p. il.

ER 1.11:ERDA-23A	052-010-00475-1	2.50

Non-Technical Summary of Distributed Solar Power, Collector Concept. 1976. 19 p. il.

ER 1.2:So 4/6	060-000-00008-5	.45

Our Prodigal Sun. Provides a brief introduction to the life and the anticipated eventual death of the sun, discusses the sun as the possible source of all future forms of energy, and talks about the problems in obtaining pure solar energy. 1974. 14 p. il.

NAS 1.19:118	033-000-00569-5	.35

Proceedings of the Solar Heating and Cooling for Buildings Workshop, 1973: Part 1, Technical Sessions. Summaries of presentations concerning solar collectors, energy storage, heating, solar air conditioning, and combined solar heating/cooling systems. Part II, "Panel Sessions," is not yet available. 1974, reprinted 1976. 226 p. il.

NS 1.2:So 4/3/pt. 1	038-000-00171-0	2.50

The Quiet Sun. A textbook of solar physics, which tries to interpret solar observations in the framework of theoretical physics. 1973. Clothbound. 344 p. il.

NAS 1.21:303	033-000-00454-1	7.55

Report and Recommendations of Solar Energy Data Workshop Held
November 29–30, 1973, and September 1974. 1974. 228 p. il.

    NS 1.2:So 4/7   003-023-00033-8 $ 2.90

Retrofitting: A Residence for Solar Heating and Cooling The Design
and Construction of the System. 1975. 94 p.

    C 13.46:892   003-003-01549-1

                      1.70

Solar Cooling for Buildings. The proceedings of a workshop on solar
cooling held February 6-8, 1974, and sponsored by the National
Science Foundation. 1974. 231 p. il.

    NS 1.2:So 4/10   038-000-00189-2   3.00

Solar Dwelling Design Concepts. 1976. 146 p. il.

    HH 1.2:So 4/3   023-000-00334-1   2.30

Solar Energy, Environmental and Resource Assessment Program,
Summary Report. 1976. 53 p. il.

    ER 1.11:ERDA 76-138   060-000-00014-0   1.15

Solar Energy for Health Care Institutions. 1976. 8 p.

    HE 20.6002:So 4   017-022-00488-7   .35

Solar Energy as a National Energy Resource. Did you know that solar
energy is received in sufficient quality to make a major contribu-
tion to the future of U.S. heat and power requirements, and there
are no technical barriers to wide application of solar energy to
meet U.S. needs? These and an astonishing number of facts concerning the potential of solar energy are contained in this report.
1974. 85 p. il.

    NS 1.2:So 4/2   038-000-00164-7   1.50

Solar Energy—Project Independence Blueprint: Final Report. This
report, compiled by the Project Independence Task Force,
analyzes the extent to which the U.S. can utilize fuel to fill our
energy needs, and the option and responsibility that will follow
the fuel use. Reprinted 1975. 584 p. il.

    FE 1.18:So 4   041-018-00012-7   6.20

Solar Energy Research and Development. Hearings before the Joint
Committee on Atomic Energy, May 7 and 8, 1974, on a bill to establish within the Energy Research and Development Administration an Office of Solar Energy Research with a 5-year, $600
million program. The hearings also discuss the present status and
the prospects of solar energy utilization and technology. 1975, reprinted 1976. 846 p. il.

    Y 4.At 7/2:So 4   052-070-02857-2   7.00

Solar Energy School Heating Augmentation Experiment. Fauquier
  High School in Warrenton, Virginia is heating five mobile class-
  rooms for about 25 cents a day. This is a report on the design,
  construction, and initial operation of their solar heating system.
  1975, reprinted 1976. 88 p. il.

NS 1.2:So 4/11   038-000-00204-0 $ 1.45

Solar Energy for Space Heating and Hot Water. 1976. 14 p. il.

ER 1.2:So 4/5   060-000-00006-9   .35

Solar Energy Utilization for Heating and Cooling. Discusses the
  technology of solar energy utilization, heat storage systems, cooling by solar energy, and other technological aspects of solar heating and cooling. 1975. 20 p. il.

NS 1.2:So 4/9   038-000-00188-4   .70

Solar Grain Drying, Progress and Potential. 1976. 13 p. il.

A 1.75:401   001-000-03632-7   .40

Solar Heating and Cooling Demonstration Program, Cycle One. A
  descriptive summary of HUD solar residence demonstrations,
  cycle one. 1976. 54 p. il.

HH 1.2:So 4/2   023-000-00338-4   1.15

Solar Heating and Cooling, An Economic Assessment. 1977. 70 p.

NS 1.2:So 4/15   038-000-00300-3   1.20

Solar Heating: Proof-of-Concept Experiment for a Public School Building. A report on a solar heating project conducted at a typical elementary school in the Baltimore County school system. From
  March 1, 1974 to May 15, 1974, a wing of the school received 90%
  of its heat from a solar heating system. That system is described
  here, along-side an analysis of its operation and recommendations
  and conclusions based upon the results of the experiment. 1974,
  reprinted 1975. 111 p. il.

NS 1.33:So 4/2   038-000-00242-2   3.05

Solar Heating, Proof-of-Concept Experiment for a Public School Building. 1975. 84 p. il.

NS 1.2:So 4/13   038-000-00242-2   2.25

Solar Power from Satellites, Hearings, Senate, Aeronautical and Space
  Science Committee, Subcommittee on Aerospace Technology and
  National Needs, 94th Congress, 2d Session, January 19, and 21,
  1976. This publication surveys concepts involving advanced aerospace technology that might help satisfy one of our greatest na-

tional needs—future sources of energy. Specifically considered are ways to collect solar power in space with satellites and to beam that power down to earth to supplement our other sources of electricity. Included too, are many ways to construct those satellites. 1976. 228 p. il.

      Y 4.Ae 8:So 4/2   052-070-03319-3 $ 2.70

Solar Thermal Energy Conversion—Program Summary, October 1976. In Preparation.   060-000-00040-9   1.45

Survey of Solar Energy Products and Services, May 1975. Solar energy is one of the leading alternative possibilities for commercial production of energy in the United States. This survey describes and explains the various solar energy research projects in private industry today, and the present conversion and use of solar energy. 1975, reprinted 1976. 545 p. il.

      Y 4.Sci 2:94-1/G   052-070-03011-9   4.60

Useable Electricity from the Sun. 1976. 8p. il.

      ER 1.2:Su 7   060-000-00010-7   .35

INDEX

A

Age, of home, effect on heat-loss, 82

Agriculture, energy use by, 8

Air conditioning, by solar power, 27–28

Air-lock vestibule, as door insulation, 54

Air system, of solar heating, problems with, 11
    on roof, 11–13
    storage of heat, 11
    on wall, 13–16

Aluminum, as reflector, 15

Angle, for collectors, 21, 91, 106–7

Appliances, energy use by, 139

Architects, dealing with, 180–83

Area, of collectors, 93–94

Attic, insulation of, 41–49, 72–73

Axis, of home, effect on solar suitability, 86–88

B

Basement, insulation of, 57–61

Batts, as insulation material, 73

Beadboard, for insulation, 40,75

Beadwall, as window insulation, 52

Bids, from contractor for work, 161–62

Blankets, of insulation material, 74

Blocbond, for window insulation, 27

Borrowing, for solar installation, 216–31

Btu's (British thermal units), as measurement of heat energy, 38

Building codes, effect on solar systems, 194–99

Buildings, solar, in United States, 154–57

C

Caulking, to reduce heat-loss, 37, 55

Cellulose fiber, for insulation, 40

Chicken wire, to hold up insulation, 58

Climatic zone map, 12

Coal, as fuel source, 8

Collectors, angle of, 21, 91, 106–7
     insulation of, 68–71
     on roof, 19
     size of, 20–21, 93–94

Concrete, for heat storage, 25

Concrete floors, insulation of, 65–68

Condition, of home, effect on heat-loss, 83

Conduction, as cause of heat-loss, 37–38

Conservation, of energy sources, 35
     family habits, 112–25

Construction, of home, effect on heat-loss, 83

Contract, with architect, 184–87
     with contractor, 165–77

Contractors, dealing with, 159–80

Controls, of solar system, problems with, 152

Conversion, of existing homes to solar, 18
     determining suitability for, 79–102

Corrosion, of solar collectors, 150

Costs, of solar heating, 30, 49

Crawl space, insulation of, 57–61

D

Dampers, manually operated, 3, 152

Deciduous trees, for summer shade, 84

Design, problems with, 150–52

Doors, insulation of, 54, 70

Drapes, as window insulation, 51–52

Ductwork, insulation of, 58, 59

E

Electrical wiring, insulating around, in attic, 46

Electricity, generated by solar cells, 28
     generated by wind power, 28–29
     use by family, 116–19

Energy use, by household appliances, 139
     reduction of, 35
     in United States, 8, 36

Equipment, for insulating attic, 43–44

Estimates, from contractor for work, 161–62

Evergreens, as windbreak, 84

F

Family evaluation test, for solar suitability, 112–25

Felt fiber strips, as weatherstripping, 55

Fiberglass covers, on collectors, problems with, 152

Fire hazard, of some insulation, 61

Floors, concrete, insulation of, 65–68

Foamboard, for insulation, 40

Foam rubber stripping, as weather stripping, 55

Fossil fuels, limits of, 8

Free-standing collectors, near house, 22–23

Furnace, insulation of, 61–62

G

Gas, as fuel source, 8

Glass fiber, for insulation, 39, 40, 64, 73

Guarantees, on contractor's work, 163

H

Heat exchangers, 99

Heating habits, of family, 112–14

Heating system, existing, effect on solar suitability, 95–96

Heat-loss, causes of, 36–38
    effect on solar suitability, 81–82

Heat storage, in air system, 11
    insulation of, 69–71
    in passive systems, 25–26

Hot water, use by family, 114–16

Hot water heater, insulation of, 62–63

I

Incentives, for solar installation, 201–3, 207–13

Industry, energy use by, 8, 36

Infiltration, as cause of heat-loss, 36, 37

Insulation, 33–71
    of attic, 41–49, 72–73
    of basement, 57–61
    of concrete floors, 65–68
    of doors, 54, 70
    of furnace, 61–62
    of hot water heater, 62–63
    of inside walls, 63–65
    materials for, 39–40, 73–76
    priorities of, 38
    R values of, 72–73
    of solar collectors, 68–71
    of windows, 26–27, 49–54, 70

Insurance, for solar systems, 205–6

L

Labor laws, effect on solar installation, 205

Landscaping, effect on home's heat-loss, 84
    effect on home's solar suitability, 100–101

Leaks, in systems, problems with, 149–50

Legal considerations, affecting solar heat, 189–206

Light-fixture housings, insulating around, in attic, 48

Loans. See Borrowing

Location, of home, effect on solar suitability, 80

Loose fill, of insulation materials, 75–76

M

Manufacturers, of solar collectors, 238–47
    of solar hot water equipment, 126–38

Materials, for insulation, 39–40

Moisture control, vapor barriers for, 41

Mortgages. See Borrowing

N

Nuclear power, as fuel source, 8

O

Oil, as fuel source, 8

Option chart, for solar conversion, 125

Orientation, of home, effect on solar suitability, 85–90

P

Paint, vapor-proof, 41

Passive systems, of solar heating, 23–26

Payments, to contractor, 178

Percentage, of heat to be supplied by solar system, 13

Performance specification, for bidders, 162

Perlite, as insulation material, 75

Pipes, insulation of, 58

Plastic foam. See Urea-formaldehyde

Plastic jugs, for heat storage, 146

Poll, of solar homeowners, 142-53

Problems, with solar systems, 148–52

Progress payment, to contractor, 178

Publications, federal, on solar energy, 249–55

R

Records, of dealings with contractor, 179–80

Recycling, of wastes, 121–23

References, of contractor, 160

Reflectors, to increase available sunlight, 15–16

Restrictions, physical, to solar installation, 97–102

Rigidboard, as insulation material, 60, 75

Rocks, for heat storage, 25

Rock wool, for insulation, 39, 40, 64, 73
Roof slope, effect on solar suitability, 90–93
   measuring of, 92
R value, definition of, 37
   of insulation, 40, 72–73

S

Shades, as window insulation, 51–52
Shutters, as window insulation, 27, 50–51
Skylight, insulation of, 54
Slope of roof, 90–93
   for solar collector, 91, 106–7
Solar cells, for generating electricity, 28
Solar heating, principles of, 9
   types of systems, 11–26
Solaris, system for solar heating, 7
Storage, of electricity, 28
   of heat, 98
     in air system, 11
     in water system, 20–21
Storm doors and windows, to reduce heat-loss, 37, 49
Style, of home, effect on heat-loss, 82–83
Suitability, of home, for solar heat, 79–109
Sunlight, available to home, 85–90
   legal access to, 191–94
Surewall, for window insulation, 27

T

Tax laws, effect on solar installation, 201–3
Thermostat, setting to conserve energy, 35
Thickness, of insulation materials, 76
Thomason, Harry, discovery of solar heat principles by, 7
Tools, for insulating attic, 43–44
Transportation, energy use by, 8, 36
Travel habits, of family, 119–21
Type, of home, effect on solar suitability, 80
Trees, as shade and windbreak, 84

U

Urea-formaldehyde, for insulation, 40, 65, 74
Urethane, for insulation, 40, 75
Utilities, effect on solar installations, 203–5

V

Vapor barriers, on insulation, 41
Variance, to zoning law for solar installation, 199
Venetian blinds, as window insulation, 53

Ventilation, of home, when properly insulated, 37

Vermiculite, for insulation, 40, 75

W

Walls, insulation of, 63–65

Waste recycling, by family, 121–23

Water, for heat storage, 25

Water heating, by solar collectors, 18–19

Water system, of solar heating, of roof, 16–21
    storage tank, sizes of, 21–22
    on wall, 21–22

Weather stripping, to reduce heat-loss, 37, 55

Withholding money, from payments to contractor, 178

Windbreaks, to reduce heat-loss, 37, 84

Windows, insulation of, 26–27, 49–54, 70

Wind power, for generating electricity, 28

Z

Zomeworks, source of window insulation, 27, 52

Zoning laws, effect on solar installations, 199–201